HANDBOOK OF
DAIRY SCIENCE

HANDBOOK OF
DAIRY SCIENCE

Edited by

Dr. Vijay D. Kele

Delve
PUBLISHING

www.delvepublishing.com

Handbook of Dairy Science

Dr. Vijay D. Kele

Delve Publishing
4164 Lakeshore Road
Burlington, ON L7L 1A4
Canada
www.delvepublishing.com
Email: orders@arclereducation.com

ISBN: 978-1-77956-361-3 (Hardcover)

ABOUT THE EDITOR

Dr. Vijay D. Kele is currently an Associate Professor in the Department of Dairy Technology and Food Technology at Parul Institute of Technology, Parul University, Vadodara, Gujarat. He holds a B.Tech in Dairy Technology, MAFSU, Nagpur, M.Sc. in Dairy Science, MBA in Food and Agribusiness Management from the University of Mysore, where he earned four (04) gold medals in Academics. He also holds a Doctorate in Dairy Extension Management / Agribusiness Management.

With 14 years of teaching and 3 years of industrial experience, Dr. Kele has published 4 international books, 6 national books, and 28 research papers in prestigious national and international journals. Additionally, he has authored 87 popular articles, 72 extension articles, delivered 14 radio talks, and participated in 2 live television talks on Doordarshan.

He has actively contributed to the academic community by writing 22 practical manuals for undergraduate courses funded by ICAR, New Delhi, and has organized numerous educational events, including 133 farmer-scientist meetings, 14 exhibitions, 103 webinars, 7 short-term training programs (STTPs), 12 workshops, and 3 faculty development programs (FDPs).

Dr. Kele's research endeavors have secured funding from ICAR, GUJCOST, GSBTM, and SSIP (Government of Gujarat), as well as R&D funding from NABARD for organizing seminars and exhibitions for farmers. He also has 2 patents to his credit.

TABLE OF CONTENTS

LIST OF FIGURES

LIST OF TABLE

PREFACE

This Handbook on Dairy Science explores the composition, characteristics, and processing procedures of dairy products. The subject matter encompasses a broad spectrum of subjects, such as the chemical makeup of milk, the function of different chemicals in dairy products, the impact of processing on product excellence, and the significance of quality control in dairy manufacturing.

This book holds significance as it addresses a specific area that has not been adequately covered in the current body of research on Dairy Science. Although there is a wide range of literature available on the subject of food chemistry, only a limited number of them expressly address the topic of dairy products. This book fills the need by delivering a comprehensive and thorough examination of the chemistry of dairy products, providing vital knowledge for researchers, students, and professionals in the industry.

The information presented in this book is urgently needed because of the increasing demand for premium dairy products. With the increasing awareness of health and environmental concerns among consumers, it is crucial for dairy farmers to have a thorough understanding of the chemistry of their products. This book equips readers with the expertise and resources necessary to create dairy products of exceptional quality that satisfy consumer demands and adhere to regulatory guidelines.

Unlike other books that concentrate on specific areas of dairy production, it takes a more holistic approach. Consequently, it becomes an invaluable asset for anybody seeking to expand their knowledge on the scientific principles underlying dairy products.

This book is intended for researchers, students, and professionals that specialize in Dairy Chemistry. Researchers will discover valuable information regarding the most recent research and advancements in the field, while students will get advantages from the complete review of Dairy Chemistry themes included in the book. This book will be beneficial for professionals in the dairy sector as it offers practical insights and instructions for enhancing the quality of dairy products.

There are multiple advantages for readers who choose to read this book. Initially, individuals will acquire a more profound comprehension of the chemical composition of dairy products, enabling them to make more knowledgeable choices regarding their manufacturing and processing. Additionally, readers will acquire knowledge regarding the most recent research and advancements in the domain of Dairy Science, ensuring they remain well-informed about current trends and technology. Readers will acquire practical knowledge and instructions for enhancing the quality of dairy products, enabling them to fulfill consumer expectations and comply with regulatory criteria.

Ultimately, the book on Dairy Science serves as a great reference for individuals seeking to enhance their understanding of the scientific principles underlying dairy products. The extensive range of Dairy Science subjects covered, the direct applicability to the increasing need for top-notch dairy goods, and the potential advantages for researchers, students, and professionals alike, render it an essential read for individuals engaged in this field. I trust that this introduction has imparted a lucid comprehension of the extent, importance, pertinence, distinctiveness, intended readership, and possible advantages of this book. I encourage you to contemplate perusing it in order to enhance your grasp of dairy science.

CHAPTER 1

INTRODUCTION TO DAIRY CHEMISTRY

Contents

Every dairy product is a complex mixture of many chemical compounds. These chemical species are commonly organized into primary categories, such as water, fat, protein, lactose, and minerals, for the purpose of compositional analysis.

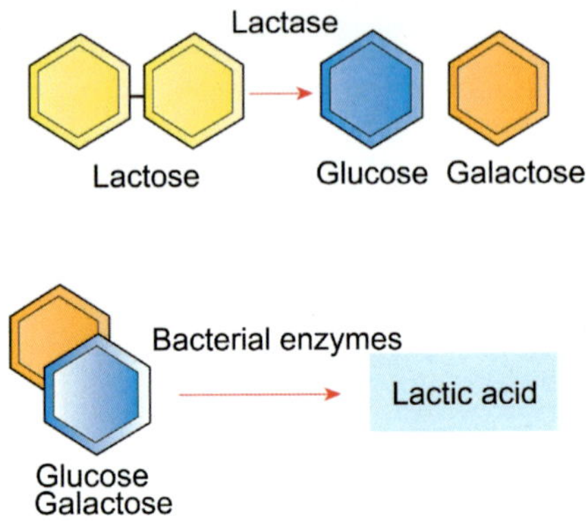

Figure 1.1. *The chemistry of milk.*

Source: Dairy Processing Handbook, 2020.

Milk is an aqueous solution containing dissolved carbohydrates, protein aggregates bound to minerals, and butterfat globules in the form of an emulsion or colloid. Every part of it facilitate□growth a□they are □pecifically de□igned to provide as nourishment for the young (Fox & McSweeney, 2007). The primary requirements include energy sources such as lipids, lactose, and protein, as well as water, essential fatty acids, vitamins, and inorganic elements. Additionally, non-essential amino acids, which are derived from proteins, are necessary for the synthesis of essential amino acids and amino groups (Figure 1.1).

1.1. HISTORICAL BACKGROUND OF DAIRY CHEMISTRY

Dairy science and technology, similar to other scientific and technological disciplines, receives assistance from several professional groups and associations that were established to advance the progress of all aspects of dairy production, ranging from economics to the supporting fundamental sciences. Dairy science is a complex network of structures that is generated from several fundamental sciences such as chemistry, biochemistry, microbiology, nutrition, physiology, and genetics (Fox, 2013). Unlike pure sciences, dairy science incorporates multiple disciplines to provide a more intricate framework. Dairy husbandry and dairy technology, the two primary branches of dairy research, largely operate separately from each other despite their clear interconnection. In the field of dairy husbandry, milk is generated on agricultural establishments, whereas in dairy technology, milk is converted into products that possess stability, health benefits, and sensory appeal.

The term "dairy science and technology" is one of numerous designations used to describe the scientific discipline that underpins dairy technology. Dairy science and technology encompasses the examination of milk's chemical makeup, the control of both beneficial and harmful microorganisms in milk and dairy products, various engineering disciplines related to milk processing, and the economic aspects of milk distribution, processing, and sales. The initial investigation in dairy science was carried out by fundamental chemists who documented their discoveries in chemical publications subsequent to the isolation and characterization of the primary constituents of milk, including lactose, lactic acid, casein, proteins, and lipids (Fox, 2013).

The implementation of food legislation, specifically the Sale of Food and Drugs Acts of 1875 and 1879 in the United Kingdom, played a significant role in advancing the study of food and dairy chemistry during its early stages. The aforementioned actions resulted in a demand for individuals known as public analysts. In response, the Society of Public Analysts was established in 1874 and began publishing an official journal called The Analyst in 1877 (Abou-Donia, 2008). The emergence of bacteriology in the late nineteenth century can be attributed to the significant contributions made by Louis Pasteur and Robert Koch.

The American Chemical Society and the Royal Society of Chemistry both have specialized divisions focused on agricultural and food chemistry. Similarly, the American Society for Microbiology and the (British) Society for General Microbiology cater to professionals specializing in the subject of food and dairy microbiology. Dairy scientists commonly retain their membership in the society that provides support for their primary field of study (Abou-Donia, 2008). The fields of dairy science and technology are closely interconnected with nutrition science due to the significant role that dairy products play in human nutrition across many regions of the globe. Dairy science and technology is a distinct field within the broader discipline of food science and technology. Institutes of food science and/or technology typically have specialized departments that cater to dairy specialists.

Following the Neolithic Revolution or the advent of agriculture, humans started consuming the milk of other mammals on a daily basis after domesticating animals. This process of development occurred spontaneously in many regions across the globe, commencing as early as 9000–7000 BC in Mesopotamia and concluding approximately 3500–3000 BC in the Americas (Abou-Donia, 2008). While domestic cattle have undergone numerous independent evolutions from wild aurochs populations over time, it was in Southwest Asia where humanity first domesticated the most important dairy animals, including goats, sheep, and cattle. The first purpose of raising animals was primarily for their meat. However, an archaeologist has suggested that the practices of dairying and utilizing domestic animals for work and obtaining hair emerged at a later stage, during a distinct revolution in secondary products that occurred in the fourth millennium BC.

The domestication of dairy animals began in Southwest Asia and then spread to South Asia (7000–5500 BC) and Europe (about 7000 BC, but they did not reach Britain and Scandinavia until after 4000 BC). Initially, the early agriculturalists engaged in the practice of extracting milk from their livestock in both Britain and central Europe. During the fourth millennium BC, European farmers migrated to the Pontic-Caspian steppe and established economies centered around animal husbandry and nomadic pastoralism. These economies primarily relied on domestic animals and their products, rather than crop cultivation (Abou-Donia, 2008). Subsequently, these economies expanded throughout a significant portion of the Eurasian steppe. Sheep and goats were introduced to Africa from Southwest Asia. However, it seems plausible that African cattle were domesticated separately, somewhere between 7000 and 6000 BC.

1.2. THE SIGNIFICANCE OF DAIRY PRODUCTS

Dairy products, milk products, or lacticinia are terms used to describe food items that are made from or contain milk. The most commonly preferred dairy animals include water buffalo, nanny goats, ewes, and cows. Dairy products encompass often seen food items available in supermarkets globally, such as cheese, yogurt, milk, and butter. A dairy is a manufacturing facility that produces dairy products. Across the globe, individuals consume dairy products to varying extents. Some individuals choose to abstain from consuming dairy products due to lactose intolerance, veganism, environmental concerns, or other health-related reasons or beliefs (Kubicová et al., 2019).

1.2.1. Health Benefits

Dairy products contribute to the development and preservation of robust bones, in addition to providing various other health advantages. The dairy group of foods offers vital nutrients that are necessary for maintaining and promoting good health in the body. Calcium, potassium, vitamin D, and protein are among the essential nutrients.

Every choice of food and beverage holds significance. Give priority to nutrition, quantity, and variety.

Calcium and vitamin D are essential minerals at every stage of life. Consuming dairy products that include these nutrients helps to enhance bone health, especially in children and teenagers during the crucial period of bone mass development (Kubicová et al., 2019).

- Promote skeletal health and prevent the onset of osteoporosis in adults.
- Low-fat and fermented dairy products have been associated with a reduced occurrence of diabetes.
- Consuming dairy products is associated with a reduced likelihood of developing gout.

Dairy products might potentially induce gastrointestinal discomfort for individuals with milk allergies or lactose intolerance. Individuals suffering from lactose intolerance commonly avoid consuming milk and other dairy products that contain lactose due to the potential occurrence of moderate adverse effects such as nausea, bloating, diarrhea, and stomach pain. These individuals may employ non-dairy milk substitutes (Kubicová et al., 2019).

Multiple reputable organizations, including the American Institute for Cancer Research (AICR), World Cancer Research Fund International (WCRF), Cancer Council Australia (CCA), and Cancer Research UK, have provided convincing evidence that consuming dairy products reduces the likelihood of developing colon cancer. According to the AICR, WCRF, CCA, and Prostate Cancer UK, there is limited but suggestive evidence that dairy products increase the likelihood of developing prostate cancer (Yerlikaya et al., 2013). According to the American Cancer Society (ACS), dairy products have the potential to decrease the risk of certain cancers while potentially increasing the risk of others. Due to this uncertainty, the ACS does not provide precise guidelines on the intake of dairy foods for cancer prevention.

Consuming dairy products that contain high levels of insulin-like growth factor 1 (IGF-1) can potentially increase the likelihood of developing cancer, particularly prostate cancer. However, a 2018 assessment found that there is not enough information to make any certain conclusions about whether consuming dietary IGF-1 is linked to a higher risk of cancer in individuals. According to the COC, it is unlikely that most consumers would absorb undamaged IGF-1 from food (Kubicová et al., 2019).

The American Medical Association (AMA) recommends that consumers replace full-fat dairy products with nonfat and low-fat alternatives. In 2017, the American Medical Association stated that there is no substantial clinical evidence connecting the eating of cheese to a reduced risk of heart disease. In 2021, Yerlikaya et al. (2013) stated that substituting full-fat dairy products with nonfat and low-fat dairy products, as well as other sources of unsaturated fat, leads to dietary patterns with higher unsaturated to saturated fat ratios. These dietary patterns are linked to improved cardiovascular health.

As of 2022, there is inadequate data to establish definitive findings, although there is a correlation between the use of dairy products, such as whole and low-fat milk, and a higher occurrence of acne. Consumption of dairy products that are low in fat and have undergone fermentation has been associated with a reduced occurrence of diabetes. Consuming dairy products is associated with a reduced likelihood of developing gout.

A study published in 2023 found a significant correlation between increased consumption of dairy products and a reduced risk of inflammatory bowel disease.

In the initial five years of life, children experience rapid growth and have high energy needs. Due to their limited stomach capacity, they must consume nutrient-rich diets in order to sustain their growth and survival. Dairy products produced with whole milk and full fat provide advantageous calories, protein, vitamins, and minerals to support the promotion of healthy growth and development (Sajdakowska et al., 2018). Milk not only provides essential nutrients for proper growth and development, but also protects teeth from tooth decay.

During adolescence, there is a significant growth of bones, with 80–90% of the skeleton being established by the age of 18. Additionally, 40–60% of the highest amount of minerals in bones is deposited during this period. Consuming a nourishing diet during adolescence might enhance bone mineral density, hence promoting robust bones in adulthood and reducing the risk of conditions such as osteoporosis (Sajdakowska et al., 2018).

Ensuring sufficient calcium intake is crucial for women during pregnancy and lactation to fulfill the nutritional requirements of both their growing kid and themselves. While it is not essential to increase calcium intake, it is imperative to consume an adequate amount to prevent calcium loss from the bones during this time.

Women above the age of 50 have an elevated risk of developing bone diseases. During menopause, women may have a decline in calcium stores in their bones, which raises their susceptibility to fractures. Multiple experiments have demonstrated that increasing milk consumption effectively reduces the incidence of bone loss in older individuals and postmenopausal women (Sajdakowska et al., 2018).

- Dairy products and milk contain a combination of nutrients that contribute to the maintenance of teeth, skin, muscles, bones, and eyesight.

- Consuming food can provide energy and alleviate feelings of exhaustion and fatigue.
- Maintain your blood pressure within a healthy level.
- Foster optimal brain development and growth.
- Enhance the functioning of the immune system.

1.3. THE CHEMICAL COMPOSITION OF MILK

Cow's milk in the United States typically consists of 87.7% water, 4.9% lactose (a carbohydrate), 3.4% fat, 3.3% protein, and 0.7% minerals, also referred to as ash. The composition of milk is influenced by various factors, including the species (such as cow, goat, and sheep), breed (such as Holstein and Jersey), the animal's nutrition, and the lactation stage (Mehta, 2015).

1.3.1. Milk Carbohydrate (Lactose)

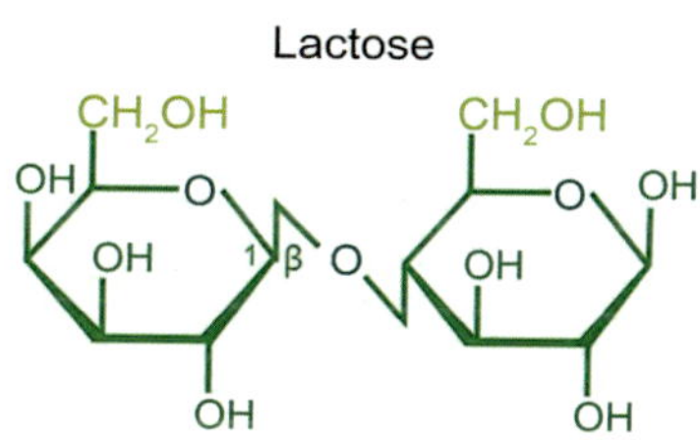

Figure 1.2. *Structure of lactose.*

Source: GeeksforGeeks, 2023.

The carbohydrate content of milk consists primarily of lactose, with minimal amounts of mono- and oligosaccharides. Lactose is a disaccharide composed of galactose and glucose. The structure of lactose is as follows (Figure 1.2):

The serum component of fluid milk solubilizes lactose. There are two forms of lactose that are dissolved in a solution: the α-anomer and the ß-anomer. These two forms have the ability to convert into one other (Mehta, 2015). The equilibrium concentration of the two forms will fluctuate depending on the temperature due to the temperature-dependent solubility of the two anomers. The equilibrium ratio at room temperature (70°F/20°C) is around 37% α-lactose and 63% ß-lactose. The α-to ß-lactose ratio is larger because the ß-anomer has lower solubility at temperatures above 200°F (93.5°C). The nutritional properties of lactose are not influenced by the type of anomer that is present (Mehta, 2015). Lactose crystallization occurs when the concentration of lactose exceeds its solubility. The physical properties of lactose crystals can significantly influence their application in culinary practices, depending on the specific crystal form. Temperature affects the equilibrium ratio between the α- and ß-lactose anomers, as stated before. α-lactose crystals are the main type of crystals formed when lactose crystals are produced at temperatures below 70°F (20°C) (Mehta, 2015).

1.3.2. Milk Fat

Fatty acid molecules are chemically bonded to glycerol, a three-carbon structure, to create fats. Triglycerides, or triacylglycerols, are the most abundant form of fat (Figure 1.3). They consist of three fatty acids that are attached to the backbone and resemble a handleless fork.

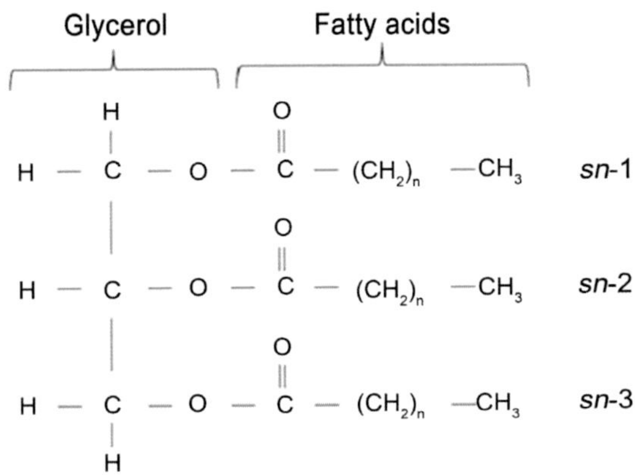

Figure 1.3. *Triacylglycerol (TAG) structure showing glycerol with three fatty acids.*

Source: Mills et al. (2017).

The fat content of milk is approximately 3.4%. Milk fat has the most intricate composition of fatty acids among all dietary fats. Mehta (2015) discovered over 400 different fatty acids in milk fat. Nevertheless, 90% of the fat found in milk is comprised of a range of 15 to 20 fatty acids. The primary types of fatty acids included in milk fat include monounsaturated fatty acids, polyunsaturated fatty acids, and straight chain fatty acids, which are saturated and consist of carbon chains ranging from four to eighteen carbons. Despite being present in trace amounts, some fatty acids contribute to the unique and attractive flavor of butter and milk fat (Mehta, 2015). The fatty acid composition of milk fat undergoes changes throughout the lactation cycle of the cow. The animal's mammary gland synthesizes fatty acids with carbon chain lengths ranging from 4 to 14. The animal synthesizes certain 16 carbon fatty acids internally, while it acquires other portions of these fatty acids from its diet. The animal's diet contains all eighteen carbon fatty acids (Fnb et al., 2009). Both the animal's energy requirements and the stage of lactation contribute to systematic fluctuations in the composition of milk fat.

1.3.3. Milk Protein

Proteins exist in diverse forms, with each having a distinct sequence of amino acids, typically consisting of hundreds of units. Protein chains can be formed by combining any of the 22 amino acids. The human body lacks the ability to synthesize nine amino acids, hence they must be obtained through dietary sources. These are commonly known as essential amino acids. Protein chains consist of amino acids that can undergo folding, resulting in the formation of three-dimensional structures through the establishment of intermolecular interactions. Proteins have the ability to arrange themselves into compact globules, adopt a relatively linear shape, or exhibit intermediate structures. The term "denatured" is used to describe the unfolding of proteins from their original chain or globular structure (Enb et al., 2009).

Milk contains 3.3% of the overall protein content. Milk proteins include all nine essential amino acids required by humans. While the mammary gland is responsible for synthesizing milk proteins, the cow's diet contributes 60% of the necessary amino acids for protein production. The genetic composition of each individual animal and the specific breed of cow have an impact on the overall quantity of protein and amino acids present in milk. Milk proteins can be classified into two primary categories based on their distinct physical and chemical properties. Phosphorus is found in the casein family, which forms a solid or settles out of a solution with a pH of 4.6. Phosphorus is not present in the serum proteins, which remain dissolved in milk at a pH of 4.6. The production of cheese curds relies on the concept of coagulation, or the generation of curds, at a low pH (Enb et al., 2009). Casein constitutes approximately 82% of the protein content in cow's milk, whereas the remaining 18% is composed of serum, sometimes known as whey protein.

1.3.4. Vitamins in Milk

Water-soluble vitamins, including B1 (thiamin), B2 (riboflavin), B3 (niacin), B5 (pantothenic acid), B6 (pyridoxine), B12 (cobalamin), C (vitamin C), and folate, can be found in milk. Milk contains vitamin B12, riboflavin, and thiamin. Niacin, pantothenic acid, vitamin B6, vitamin C, and folate are present in milk in small quantities. However, milk is not considered a significant source of these vitamins in the diet (Enb et al., 2009). Milk contains the fat-soluble vitamins A, D, E, and K. The fat content of dairy products directly influences the concentration of fat-soluble vitamins they contain. In order for reduced fat (2% fat), low fat (1% fat), and skim milk to possess equivalent nutritional value as whole milk, it is necessary to fortify them with vitamin A. Fortifying all milk with vitamin D is discretionary. Milk contains trace levels of vitamins E and K; however, it is not considered a significant dietary source of these nutrients (Enb et al., 2009).

1.4. MILK'S PHYSICAL PROPERTIES

Milk is a diluted emulsion consisting of an aqueous colloidal continuous phase and an oil/fat dispersion phase. The physical properties of milk are similar to those of water, although they are affected by the extent of dispersion of the colloidal and emulsified components, as well as the presence of various solutes (salts, lactose, and proteins) in the continuous phase. Milk possesses various physical attributes, such as density, conductivity, thermal properties, rheological behavior, redox properties, colligative properties, surface activity, buffering capacity, and color (Fox et al., 2015).

1.4.1. Density

Milk and milk products' density can be utilized:

- For converting mass to volume and vice versa.
- To calculate supplementary physical characteristics, such as kinematic viscosity, in order to ascertain the solids content.

Den□ity i□determined by dividing the ma□□of a □pecific quantity of material by it□volume. It i□influenced by the following variable□:

- The current temperature during the measurement;
- The historical warmth of the material;
- The composition of the material, namely its fat content; and
- The presence of air, which might be an issue with thicker substances.

1.4.2. Viscosity

The viscosity of milk and milk products is essential in determining the following factors:

- The rates of mass and heat transfer;
- The rate at which creaming occurs; and
- The flow condition□in dairy operation□

Except for cooled raw milk, both milk and skim milk exhibit Newtonian behavior, indicating that their viscosity is not affected by the shear rate (Fox et al., 2015). The viscosity of these items is determined by the following factors:

- **Temperature:** The voluminosity of casein micelles increases at lower temperatures, leading to an enhancement in viscosity. Whey proteins undergo denaturation, resulting in an increase in viscosity when exposed to temperatures over 65°C.
- **pH:** The voluminosity of casein micelles rises when the pH of milk varies.

Raw milk and cream exhibit non-Newtonian behavior when they are cooled, meaning that their viscosity is affected by the rate at which they are sheared. Agitation can cause partial churning or partial coalescence of fat globules, leading to an increase in viscosity (Fox et al., 2015). Turbulence can disperse fat droplets that have experienced cold agglutination, leading to a decrease in viscosity.

1.4.3. Freezing Point

Freezing point depression is a property that is determined by the molarity of the solutes, rather than the percentage by weight or volume. The freezing point of milk is primarily utilized in the dairy business for determining the quantity of water added. However, it can also be employed to ascertain the amount of lactose present in milk, estimate the quantity of whey powder in skim milk powder, and evaluate the water activity of cheese. The freezing point of milk is usually between the range of –0.512 to –0.550°C, with an average of –0.522°C (Fox et al., 2015).

A comprehensive understanding of the factors that affect freezing point depression is essential for precise analysis of freezing point data in relation to the presence of additional water. The primary considerations for interpreting freezing values for determining additional water are the nutritional health of the herd and the availability of water. In□ufficient feeding lead□to higher freezing points.

According to Luccy (2002), milk has the same osmotic pressure as blood, which means that eating large amounts of water might cause temporary rises in the freezing point. The primary factors contributing to the inadvertent introduction of water into milk are the residual rinse water and condensation present in the milking equipment.

1.4.4. Acid-Base Balance

Milk's acidity is assessed through the use of both pH and titratable acidity. Milk's pH at a temperature of 25°C often falls within the narrow range of 6.5 to 6.7. The titratable acidity of herd milks normally falls within the range of 13 to 20 mmol/L (Lucey, 2002). The measurement of titratable acidity is not widely practicable due to its significant inherent variability. However, it can be useful for monitoring changes in acidity, particularly during lactic fermentation. In such cases, pH is a more precise and reliable measurement.

Milk has several components that exhibit a buffering effect. Phosphate and caseins are the primary constituents in milk that function as buffers.

1.4.5. Properties of Optics

Various rapid, indirect assessment methods, such as proximal analysis utilizing infrared absorption or light scattering, rely on optical properties. The appearance of milk and milk products is also influenced by their optical properties. The turbidity and opacity of milk are a result of the scattering of light by fat globules and casein micelles.

Light scattering occurs when the wavelengths of the particle and the light wave are almost the same. Smaller particles cause shorter wavelengths of light to scatter. Skim milk appears slightly blue due to the fact that casein micelles disperse shorter wavelengths of visible light, such as blue, more than longer wavelength□like red. The milky color i□attributed to ß-carotene, a carotenoid compound that serves as a precursor to vitamin A and is present in the fat content of milk. The greenish tint of whey i□derived by riboflavin (Lucey, 2002).

1.5. NUTRITIONAL COMPOSITION OF DAIRY PRODUCTS

Milk is considered one of the most nutritious drinks in the world. It is an excellent source of calcium, vitamin B12, riboflavin, and various other vitamins and minerals, in addition to having a high content of high-quality protein.

Milk is an excellent reservoir of essential vitamins and minerals. Studies have shown that including milk and dairy products in a healthy diet can boost bone health and enhance the overall nutritional value of the diet (Werner et al., 2014).

However, there i□a □ignificant amount of inaccurate information circulating about the health and nutritional value of dairy products. Consequently, several individuals unfortunately opt to exclude dairy products, thereby forfeiting the manifold nutritional and health benefit□a□□ociated with dairy.

Dairy is a crucial element of a nutritious diet. Milk and dairy products are commonly disregarded as sources of calcium, magnesium, selenium, riboflavin, vitamin B12, and pantothenic acid (Werne et al., 2014). The use of dairy products ha□the capacity to offer □ub□tantial nutritional benefit□to large population□in both developed and developing countries. Most countries recommend taking at least one portion of milk or dairy products daily because of the substantial nutritional advantages they provide. Some recommend consuming up to four servings per day (Figure 1.4).

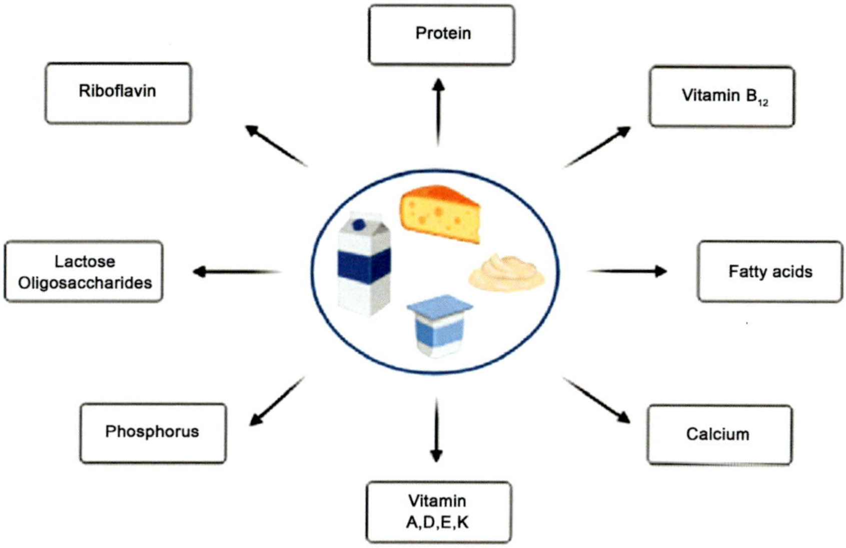

Figure 1.4. *The nutritional composition of dairy products.*

Source: Kocyigit, (2024).

1.5.1. The Nutritional Composition of Different Dairy Products

- **Dairy Milk in Liquid Form:** Milk serves as the primary source of nourishment for the offspring of animals, including humans, and has been ingested by them for thousands of years. In the United States, milk is commonly fortified with vitamin D to enhance the body's absorption of calcium and promote optimal bone health (Werner et al., 2014). There is a wide range of milk options, including skim, 2%, 1%, and whole milk.
  - *Flavored Milk:* Flavored milks, such as chocolate milk, are commonly consumed post-workout due to their combination of protein and carbohydrates, which aid in replenishing energy stores. Flavored milk can be produced using either natural or artificial tastes, and is often sweetened with sugar or other sweeteners. In recent years, there has been a noticeable change in consumer preference towards flavored milk products that have reduced or no sugar content.
- ***Cream:*** Upon standing, milk undergoes a process where cream, which is essentially fat, ascends to the uppermost layer. Cream is

available in various forms, including heavy, light, and half-and-half. Cream is commonly utilized in recipes for soups, baked dishes, and sauces due to its ability to enhance richness and flavor.

- ***Butter:*** Butter contains small amounts of other vitamins and minerals and serves as a natural source of vitamin A, which is crucial for maintaining good eye health. Butter is a versatile ingredient that may be used to bread or toast and utilized in the process of baking and cooking.
- ***Dehydrated Milk Fat:*** The water and nonfat particles are removed from milk fat using distillation, resulting in an anhydrous state. Many processed foods, such as cookies, crackers, and snack foods, contain anhydrous milk fat. The formulation for infant formula also includes anhydrous milk fat.
- ***Ghee:*** Clarified butter, often known as ghee, is a fundamental ingredient in Middle Eastern and Indian cuisine. Ghee is ideal for high-temperature cooking due to its elevated smoke point and distinctive nutty taste. Ghee is utilized in Ayurvedic medicine due to its purported therapeutic advantages.
- **Products That Have Undergone Lacto-Fermentation:** Probiotics, which are beneficial microorganisms included in fermented dairy products such as yoghurt and kefir, can assist in enhancing digestive well-being. According to Barłowska et al. (2011), Greek yogurt and other fermented dairy products provide a higher amount of protein compared to normal yogurt. Non-dairy components such as soy or coconut milk can be used to produce fermented dairy products.
- ***Cheese:*** The production of cheese entails the coagulation of milk proteins, separation of whey, and aging of curds. Cheese is available in a diverse range, with each type having a unique taste and consistency. Cheese contains significant amounts of saturated fat and sodium, but it is also an excellent source of calcium and protein.
- ***Ice Cream:*** Ice cream is a frozen dessert made from materials such as milk, cream, sugar, and flavorings. The first recorded instance of ice cream being served in the United States was in Maryland in 1744.

Ice cream is widely enjoyed across the globe, with each country offering its own unique flavors and types.

- **Evaporated Milk:** The advent of evaporated milk in the late 1800s enabled the preservation of milk without the need for refrigeration. Evaporated milk, being more viscous and dense compared to regular milk, is commonly utilized in recipes that require cream or whole milk. Due to its similar flavor and texture to cream, evaporated milk can also serve as a coffee creamer.

- **Products Derived from Dehydrated Milk:** During World War II, powdered milk was created to provide soldiers with a convenient and long-lasting source of milk. Dehydrated milk products are a convenient essential for the pantry due to their ability to remain stable at ambient temperature for extended durations. An alternative to fresh milk can be created by reconstituting powdered milk with water (Barłowska et al., 2011).

Probiotics are viable bacterial cultures with unique characteristics that allow them to thrive in the digestive system and are found in some fermented foods. They have health advantages and can assist in maintaining the natural equilibrium of the gut microbiota when consumed in enough amounts. Yogurt and cultured drinks are widely consumed dairy products that include probiotics, making them highly popular and easily available (Barłowska et al., 2011). These products are ideal for delivering probiotics because the presence of fat, protein, and carbs in milk forms a protective matrix that enables the bacteria to survive in both the gut and the harsh conditions of the stomach.

1.6. DAIRY INDUSTRY PROCESSING TECHNIQUES

Pasteurization is the first step in the process of treating milk. To achieve pasteurization of milk or milk products, it is necessary to subject each individual particle to a specific temperature for a defined duration (63°C for 30 minutes). By undertaking this process, the elimination of germs and other bacteria that may pose a threat to the well-being of clients is ensured.

The weighbridge is utilized for the transportation of milk to the dairy processing facility, where its weight is automatically documented. Simultaneously, data is transmitted wirelessly from an onboard computer to a data capture system, where it is stored as records of temperature and milk volume for each farm. The temperature should range from 4 to 6°C (Smit, 2003). Each supplier routinely gathers milk samples in aseptic containers at the time of purchase, which are then handed over to a laboratory technician for thorough analysis. Milk that deviates from conventional milk in terms of composition, flavor, and aroma has a lower quality grade. To reduce milk sloshing, the technician additionally gathers a composite sample from every compartment in the refrigerated truck.

Once the initial analytical tests are completed, the raw milk in the milk container truck is moved to the whole milk intake bays. At this stage, the driver joins the milk hoses. The milk is transferred into silos, which are expansive storage containers with a minimum capacity of 300,000 liters (Smit, 2003). In order to initiate the process of transferring milk into all of the silos, the driver has two options. They can either utilize a key fob, which is a passive wireless electronic device that commonly employs radio frequency ID technology, or they can enter the truck's identification number into the control panel of the pump. During the process of being transferred into the silo, the milk is automatically cooled to a temperature range of 4–6°C utilizing a heat plate exchanger. Monitoring of the processing can be used to track the time it takes to connect and disconnect the milk intake hose during offloading and setup (Smit, 2003).

A reliable measure of performance is the pace at which the fluid flows, and the duration of pumping can differ, indicating the distinct performance characteristics of each pump. Factors such as the milk quantity in the receiving silo, the number of bends and valves in each pipeline, and the resulting variations in backpressure are significant components that might impact the duration of pumping (Smit, 2003). To optimize the milk offload process, it is important to ensure that the pumps are functioning effectively and efficiently. Additionally, intelligent time slot management can be employed to schedule the truck supply.

1.6.1. Centrifugation, Separation, and Clarity

Each milk processing facility has its own distinct set of methods. To ensure the protection of the machinery used in downstream processing, it is necessary to clarify milk upon its arrival at the dairy. This procedure involves removing impurities such as dirt particles, including sand, soil, dust, and precipitated protein. Moreover, milk can be purified by employing centrifugation and microfiltration techniques to exclude bacteria, spores, and somatic cells.

Leukocytes and other somatic cells are removed, reducing the quantity of Listeria that is enclosed within the leukocyte (Smit, 2003). At this point, reducing the number of microorganisms can reduce the mass of biofilms, hence enhancing the efficacy of the HPE (Smit, 2003). In addition, the process of bacterial milk clarifying not only extends the shelf life of dairy products, but also improves their sensory characteristics, while simultaneously eliminating any complications that may arise during the age of cheese. Clarifiers and milk separators have distinct roles, while being categorized as centrifugal separators. While any centrifuge can function as a clarifier, it is often the ones with a significant hydraulic capacity that are commonly used for this purpose. The clarifier can utilize milk that is either hot (50–60°C) or cold (below 8°C).

Centrifuges are mostly employed in dairy processing facilities to separate heated milk. The objective is to extract the serum, often known as skim milk, from the globular milk fat. This process is referred to as skimming. This process is commonly combined with an in-line fat standardization system for both milk and cream, and incorporated into the pasteurization line. Separation usually takes place within the temperature range of 122 to 140°F (50 to 60°C) (Munir et al., 2015). The cream separator can be adjusted to control the fat content of the expelled cream within a range of 20 to 70%. In the dairy industry, the phrases "separation" are used to describe the ongoing process of using centrifugal force to separate solid particles, known as a "clarifier," as well as the separation of cream using a device called a "separator" or "bactofuge."

The microbiological integrity of milk powders is of utmost significance. Microfiltration or bactofugation can effectively eradicate 99.9% of the spore-forming bacteria during the initial stage of processing, prior to heat treatment. Standardization refers to the procedure of increasing or decreasing the quantities of fat and solids-not-fat (SNF) in milk.

The process of supplying milk to the consumer market and using it to manufacture various dairy products, such as ice cream, cheese, condensed milk, and milk powder, is often practiced (Munir et al., 2015). Standardization is typically performed to guarantee a uniform milk fat content in the final dairy product.

1.6.2. Heating Up

The primary objective of pasteurization was to regulate the tuberculosis (TB) caused by *Mycobacterium bovis*. Nevertheless, due to the annual screening of cows for TB and the culling of positive cases from herds, TB is no longer a concern. *Coxiella burnetii*, the bacterium responsible for Q fever in humans, necessitates pasteurization at a temperature of 161°F (71.7°C) for a duration of 15 seconds. Although the TB bacillus is extremely resistant to heat, the current accepted practice for pasteurizing milk is to heat it to 63°C (145°F) for a duration of 30 minutes (Munir et al., 2015). Heat processing can lead to protein denaturation, loss of vitamins and natural antioxidants, as well as a decrease in texture, freshness, and delicate fragrance and flavor components.

1.6.3. Desiccation and Vaporization

Milk powder is produced by spray-drying pre condensed milk. A falling film evaporator is commonly employed in the dairy industry to concentrate milk from around 13% total solids (TS) to a target of up to 52%. Evaporation refers to the process of removing a solvent from a solution or slurry (Munir et al., 2015). Milk is a colloid where water acts as the solvent. Freeze-drying is an alternative method for removing water.

1.7. PROSPECTS FOR FUTURE DEVELOPMENTS IN DAIRY CHEMISTRY

The future developments in dairy chemistry are centered around exploring novel uses for fractionation procedures and separation techniques. The techniques mentioned are chromatographic separation, high-pressure processing, high-pressure homogenization, enzymatic crosslinking using transglutaminase, ultrasonic processing, simultaneous emulsification, and mixing (Gökırmaklı & Bayram, 2017).

Consumer tastes are increasingly favoring plant-based dairy substitutes, including non-dairy milk, plant-based yogurt, and plant-based cheese. The growth in diversity has resulted in a higher consumption of these alternatives.

Consumers are increasingly adopting a proactive approach to avert health issues. This has intensified their preference for wellness and personal health products. Dairy products are recognized for their diverse health advantages, such as promoting skin and brain health, as well as supporting joint health (Gökırmaklı & Bayram, 2017).

Consumers prioritize holistic health, encompassing mental wellness, as a fundamental priority. They have specific concerns with depression, exhaustion, and anxiety. Researchers have been working on creating dairy products that offer mental health benefits. For example, they have used vitamin B6 to combat weariness and sleepiness, as well as valerian to promote sleep and relaxation.

There is an increasingly recognized connection between gut and digestive health and overall holistic well-being. Consumers have a keen interest in acquiring further knowledge about the fermentation process employed in dairy products. The newly launched yogurt in Australia has probiotic strains to enhance immunity, prebiotic dietary fiber to promote digestive health, and calcium.

Cow methane emissions pose a substantial issue within the dairy sector. Multiple products and feed additives are now being created with the aim of diminishing methane production in cows. These supplements include of seaweed, grains treated with anti-methane compounds, yeast enhanced with bromoform, garlic, probiotics, and feed additives including essential oils.

Wearable sensors are being increasingly adopted on dairy farms worldwide. Developers are currently working on the creation of advanced GPS/Lora tags that possess enhanced affordability, reduced weight, and increased power. The utilization of these tags can aid in the identification of troublesome cows and facilitate the management of grazing activities (Gökırmaklı & Bayram, 2017).

Agscent, a startup, is now working on the development of a digital breathalyzer specifically designed for diagnosing dairy-related issues. This device utilizes breath analysis to forecast a range of health issues in cows and cattle, including illnesses and pregnancy. The utilization of this technology has the capacity to fundamentally transform the process of identifying issues on expansive agricultural estates.

The dairy business is also employing artificial intelligence. According to Gökırmaklı and Bayram (2017), cameras that have artificial intelligence (AI) capabilities can accurately assess the physical condition and movement abilities of cows. This technology can be helpful in managing work on farms.

In summary, the future directions in dairy chemistry primarily involve the advancement of novel separation methods, catering to consumer demands for plant-based alternatives, addressing comprehensive health considerations, mitigating methane emissions, employing wearable sensors and diagnostic tools, and integrating artificial intelligence to enhance farm management (Gökırmaklı & Bayram, 2017).

CHAPTER 2

COMPONENTS OF MILK

Contents

Milk is a complex biological fluid including carbs, proteins, minerals, vitamins, lipids, and enzymes. Mammalian milk universally comprises water, proteins, lipids, carbs, vitamins, and minerals. Nevertheless, the composition of these components varies significantly between ruminants and non-ruminants (German et al., 2002). Milk composition (Figure 2.1) exhibits significant variability due to genetic, physiological, nutritional, milking frequency, and environmental factors. The composition of milk undergoes the most significant changes during lactation. The initial milk produced after calving, known as colostrum, has a reduced lactose level and an elevated fat and protein content, especially immunoglobulin. The milk production increases throughout the initial stages of lactation and subsequently declines progressively towards the conclusion of the process after reaching its maximum level. Mastitis causes changes in the composition of milk, resulting in increased levels of potassium and decreased levels of lactose (German et al., 2002).

Milk contains a substantial amount of bacteria and somatic cells. Typically, milk from animals that consume highly nutritious feeds contains higher levels of fat and protein compared to milk from animals that use less nutritious feeds. There is a direct relationship between the fat and protein contents of milk in the dairy cattle herd. However, the average levels of these components vary among different cattle breeds (German et al., 2002).

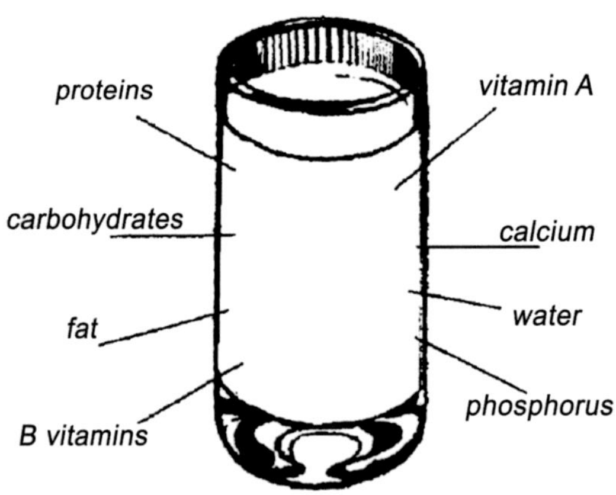

Figure 2.1. *Main components of milk nutritionally.*

Source: Zewdie, (2015).

2.1. WATER CONTENT IN MILK

The primary constituent of milk is water. The water content of milk produced by dairy cows constitutes approximately 87%, surpassing the proportion of any other substance. The other constituents of milk are dissolved, dispersed as colloids, and emulsified in water (Büttel et al., 2008).

The remaining thirteen percent consists of various compounds such as proteins, lipids, lactose, vitamins, and minerals. The water content of milk is crucial for maintaining its liquid state and plays a significant role in the body's processes of hydration and delivery of nutrients.

The milk produced by different varieties of buffaloes and cows contains around 83.18% water. The percentage range for commercial milk is 83.18 to 86.50, as stated by Büttel et al. (2008). A minuscule portion of the water in milk undergoes hydration with the salts and lactose, whereas a substantial portion is bound by proteins.

Milk is an effective natural hydrator since it contains a significant amount of water. Milk contains electrolytes such as sodium, potassium, and magnesium, which help maintain the body's fluid balance. Overall, milk is an excellent provider of nutrition and hydration.

Through the utilization of filtration and centrifugal processes, it was observed that the hydration of cheese curd was affected by variations in temperature. Specifically, an increase in temperature resulted in a decrease in the hydration of casein in both acidified and naturally soured milk. "Bound" water refers to the water molecules that are tightly associated with the milk solids in a solution. This association can occur in solutions with milk solids concentrations of up to 65 grams per 100 grams of water. Typically, as individuals get older, their casein content becomes more hydrated (Büttel et al., 2008). Previously cooked milk resulted in greater curd amounts compared to raw milk. It was shown that as the temperature increased, the hydration of albumin reduced. However, this effect was less pronounced compared to casein. Each salt made an effort to influence the hydration of the casein. Whether there was an increase or reduction depends on the location of the solution pit.

During the study of cheddar cheese, researchers noticed that the addition of salt to the curd, at any point in the process, leads to a significant decrease in the lactose content of the whey. This suggests that the bound water is being released. After the curd is sliced, lactose becomes more concentrated in the whey compared to its original concentration in the milk. Based on their calculations, approximately 2% of the water in milk is bound, indicating that 80% of the casein in cheese curd consists of bound water. Additional studies and corroborating results indicate that the curd has a water content of 0.2–0.3 grams per gram of dry matter (Büttel et al., 2008).

The amount of water available for dissolving sucrose is 100 grams less the quantity that is "bound." A significant decrease in the freezing point suggests that the molar concentration of sugar has exceeded the expected level (Büttel et al., 2008). The measurement of bound water, which refers to the amount of water that is kept and not available for the sucrose solution, is determined by the magnitude of the excess depression.

The amount of water bound is influenced by the overall solids content, as well as factors such as acidity, salt balance, and protein stability, whether they are inherent or acquired from previous treatment. Consequently, the amount of water that is tightly held inside the structure of a substance may vary between two samples that have the same amount of solid particles. The quantification of the water content that is chemically attached to molecules in milk, skim milk, cream, buttermilk, condensed skim milk, and colostrum milk is determined (Büttel et al., 2008).

2.2. MILK PROTEINS

The milk has a protein level of 3.3%. Milk proteins include all nine essential amino acids required by humans. The mammary gland is responsible for synthesizing milk proteins; however, the cow's diet contributes 60% of the necessary amino acids for protein production (Goulding et al., 2020). The genetic composition of each individual animal and the specific breed of cow have an impact on the overall quantity of protein and amino acids present in milk.

Milk proteins can be classified into two primary categories based on their distinct physical and chemical properties.

Phosphorus is found in the casein family, which forms a solid or settles out of a solution with a pH of 4.6. Phosphorus is not present in the serum proteins, which remain dissolved in milk at a pH of 4.6. The protein composition of cow's milk consists of around 82% casein and 18% whey protein. The casein protein family consists of various forms of casein (α-s1, α-s2, ß, and 6), each with distinct functional features, genetic variations, and amino acid content (Goulding et al., 2020). Milk contains micelles, which are compounds that suspend caseins. Caseins possess an open and extremely unpredictable structure due to their elevated proline concentration and changeable composition of amino acids. The high phosphate content in the casein family allows it to form calcium phosphate ions by binding to calcium. Casein proteins in milk provide a substantial amount of calcium due to their high concentration of phosphate. This allows milk to contain a larger amount of calcium than if it were completely dissolved in solution (Goulding, Fox, & O'Mahony, 2020). The 6-casein consists of a carbohydrate segment that is attached to the protein chain and is located on the outer surface of the casein micelle. Phenylalanine is absent in the protein fragment that is formed when 6-casein is broken down into particular amino acids during the cheese-making process. The milk glycomacropeptide section offers phenylketonurics a unique protein supply (Goulding et al., 2020).

The serum protein family consists of approximately 50% ß-lacto globulin, 20% α-lactalbumin, as well as blood serum albumin, immunoglobulins, lactoferrin, transferrin, and various other tiny proteins and enzymes. Every whey protein possesses a distinct composition and a spectrum of variances, similar to the other primary components found in milk. Whey proteins lack phosphorus, however they do include a considerable amount of sulfur-containing amino acids. These induce the protein to establish disulfide bonds, which confer the chain with a condensed, globular configuration. Denaturation is the phenomenon in which the compact structure of a substance is lost as a result of the disruption of disulfide bonds. Denaturation is advantageous in yogurt production as it improves the protein's ability to bind water, hence enhancing the texture of the yogurt.

Several proteins in milk have ambiguous functions and may potentially be byproducts of milk synthesis rather than serving a specific purpose. ß-lactoglobulin is thought to function as a carrier for vitamin A. It is worth mentioning that human milk does not include ß-lactoglobulin. α-lactalbumin has a crucial role in lactose production within the mammary gland (Goulding et al., 2020).

The impact of immunoglobulins on animals' immune systems remains uncertain, as it is unknown if their effects are analogous to those observed in humans. Cow's milk is being considered as a potential source of lactoferrin for commercial purposes due to its importance in iron absorption, together with transferrin.

2.2.1. Characteristics of Milk Protein's Physical Properties

Micelles are formed when the caseins in milk's aqueous phase mix together. The casein micelles consist of subunits of different caseins (ß, α-s1, and α-s2) that are connected by calcium phosphate bridges within. The micelle is stabilized in solution by a coating of 6-casein, as described by (Fox et al., 2015).

When comparing fat globules in homogenized milk, which are around 1 μm in diameter, to casein micelles, the micelles are spherical and have a diameter ranging from 0.04 to 0.3 μm. Due to the permeability of the casein micelles, the water phase can freely enter and exit the micelle without any limitations. Casein micelles are continuously changing, yet remain stable structures that stay dissolved in a solution. They can undergo dehydration and subsequent rehydration without experiencing any adverse effects, and they can be subjected to boiling or cooling. Temperature fluctuations induce the migration of ß-casein and certain calcium phosphate molecules into and out of the micelle structure.

However, these variations do not impact the nutritional properties of the protein or minerals. Whey proteins are individual entities that are dissolved in the liquid phase of milk (Fox et al., 2015).

2.2.2. Deterioration of Milk Protein

Proteins can undergo degradation by photolysis or enzymatic hydrolysis. Proteases are the primary enzymes accountable for the degradation of proteins. Milk proteases can originate from different sources, such as bacterial contamination in the air, intentionally introduced bacteria for fermentation, and somatic cells present in milk. In certain instances, proteases can yield advantageous outcomes, as observed in the manufacturing of yogurt and cheese. Consequently, in these scenarios, bacteria possessing desirable proteolytic characteristics are introduced to the milk.

Milk that experiences undesired degradation, specifically proteolysis, exhibits an unpleasant flavor and diminished quality. Plasmin is the primary protease in milk that plays a crucial role in the manufacture of cheese. It initiates the process of proteolysis during ripening, which is responsible for the development of the desirable flavors and textures in cheese (Fox et al., 2015).

Milk contains two light-sensitive amino acids, methionine and cystine, which can degrade when exposed to light. Consequently, the milk's natural flavor and nutritional content are diminished due to the absence of these two amino acids.

2.2.3. The Impact of Heat Treatment on Milk Proteins

The caseins remain unaffected by heat treatment. The nutritional and functional properties of casein proteins will not be altered by typical high temperature short time (HTST) pasteurization settings. Casein and whey proteins can undergo interactions during high temperature processing, which can impact their functional attributes while leaving their nutritional properties unaffected. ß-lactoglobulin has the ability to form a protective layer around the casein micelle when exposed to high temperatures, which helps to prevent the curdling of cheese (Fox et al., 2015). Whey proteins are more susceptible to heat compared to caseins. HTST pasteurization does not alter the nutritional and functional properties of proteins. High temperatures can cause the denaturation of ß-lactoglobulin, which is advantageous for the production of specific meals (such as yogurt) and ingredients since its proteins have the ability to bind water. Proteins experience denaturation, a process that alters their physical structure but often does not significantly impact the composition of amino acids or their nutritional worth. Intense heat treatments, such as ultra-high pasteurization, can potentially diminish the nutritious content of milk and damage amino acids that are sensitive to heat. In contrast, α-lactalbumin, which is a kind of whey protein, exhibits exceptional thermal stability.

2.3. MILK CARBOHYDRATES

The carbohydrate content in milk is primarily composed of lactose, accounting for 4.9%, with minimal amounts of mono- and oligosaccharides. Lactose is a disaccharide composed of galactose and glucose. Each cup of milk typically contains a carbohydrate content ranging from 11.3 to 11.9 grams (Crisà, 2013). Due to their higher fat content by weight compared to ordinary or skim milk, both heavy cream and half & half actually have a lower carbohydrate level per serving than an equivalent amount of the latter. Consequently, those adhering to a low-carbohydrate diet frequently like these foods.

Milk contains the highest amount of carbohydrates compared to other dairy products. Full-cream milk contains the highest quantity of carbs per cup or glass, with 11.3 grams. Low-fat milk (2%) contains 11.7 grams of carbs, while fat-free or skimmed milk contains 11.9 grams (Crisà, 2013).

Buttermilk has an extra 11.7 grams of carbs. Cream possesses a lower number of energy levels compared to other dietary items, and full-cream contains approximately 6.6 grams of carbs. According to Crisà (2013), a cream with decreased fat content, known as "light" cream, has a slightly higher carbohydrate content compared to ordinary cream. Goat milk and soy milk are two alternative types of milk that should be noted. Soy milk contains only 4 grams of carbs per cup, whereas goat's milk contains 10 grams.

Yogurt, containing approximately 8 grams of carbohydrates, is an exemplar of a dairy product with substantial carbohydrate content. The carbohydrate content of yoghurt starts off at the same level as milk, however it might decrease during the fermentation process.

Nevertheless, certain dairy products may not be conducive to good health for individuals. The following examples are characterized by their high caloric density and high fat content, primarily due to the significant quantities of added sugar:

A can of evaporated milk has 25.3 grams of carbs and a substantial amount of fat. Low- or non-fat evaporated milk has a low-fat level of 29 grams and a high carbohydrate content, as reported by Crisà in 2013.

2.3.1. Physical Properties of Lactose

The serum component of fluid milk solubilizes lactose. There are two forms of lactose that are dissolved in a solution: the α-anomer and the ß-anomer, which have the ability to transition between one other. The equilibrium concentration of the two forms will fluctuate based on the temperature due to the temperature-dependent solubility of the two anomers. The equilibrium ratio at room temperature (70°F/20°C) is around 37% α-lactose and 63% ß-lactose. The α-to ß-lactose ratio is larger because the ß-anomer has lower solubility at temperatures above 200°F (93.5°C). The nutritional properties of lactose remain unchanged regardless of the specific anomer present (Crisà, 2013).

Lactose crystallization occurs when the concentration of lactose exceeds its solubility. The physical properties of lactose crystals can significantly influence their application in culinary contexts, depending on the specific crystal type. The equilibrium ratio between the α- and ß-lactose anomers is influenced by temperature. α-lactose crystals are the main type of crystals formed when lactose crystals are produced at temperatures below 70°F (20°C). The α-monohydrate lactose crystals are generated with exceptional hardness.

Gums are commonly employed in ice cream to inhibit the crystallization of lactose. ß-lactose is sometimes preferred in some bakery applications due to its increased sweetness and solubility compared to the α-monohydrate lactose.

The lactose solution undergoes rapid dehydration to prevent crystallization and transforms into an amorphous material with a glassy appearance. Milk powders contain lactose glass, which is responsible for the formation of clumps. The intended outcome of the clumping process is a milk powder that readily dissolves in water.

2.3.2. The Impact of Heat Treatments on the Properties of Lactose

The conventional pasteurization settings do not have a substantial impact on the lactose level of fluid milk. Browning and isomerization reactions can take place at elevated temperatures necessary for spray drying and ultra-high temperature (UHT) pasteurization of products with extended shelf life. These methods can influence the nutritional characteristics and overall quality of the food. The browning process, scientifically referred to as the Maillard reaction, occurs when lactose and protein in milk mix, resulting in undesirable odors and colors. Additionally, this reaction leads to a decrease in the availability of the amino acid lysine in milk protein (Ma et al., 2014). The isomerization reaction transforms lactose into lactulose through a chemical rearrangement. Cellulose is produced by the pharmaceutical business for the purpose of making tablets.

Lactose, also referred to as milk sugar, is present in all types of milk. Lactose, a type of simple carbohydrate, makes up around 2% to 8% of the total weight of milk. Simple carbohydrates are the most easily metabolized by the body due to their uncomplicated composition. While simple carbohydrates often offer a rapid energy surge to the body, they are not as nutritionally dense as complex carbohydrates. Simple carbohydrates, such as lactose, sucrose, fructose, and glucose, can be found in various sources. Fructose is present in fruits like pears and grapes, while glucose is the sugar that your body converts all other carbohydrates into before utilizing them (Ma et al., 2014).

Simple carbohydrates, such as lactose found in milk, are occasionally categorized alongside refined simple carbohydrates, such those in table sugar or white flour, and referred to as "unfavorable" carbohydrates. However, unless you adhere completely to a low-carb diet, unrefined simple carbohydrates are generally more beneficial and contain higher nutritional content compared to refined simple carbohydrates, which are commonly present in processed foods. Therefore, they should not be automatically avoided.

2.4. FATS IN MILK

2.4.1. The Chemical Composition and Properties of Milk Fat

The fat content of milk is approximately 3.4%. Milk fat has the most intricate composition of fatty acids among all dietary fats. Over 400 different fatty acids have been identified in milk fat. Nevertheless, over 90% of the fat found in milk is made up of a combination of 15 to 20 fatty acids. The primary fatty acids included in milk fat include monounsaturated fatty acids (16:1, 18:1), polyunsaturated fatty acids (18:2, 18:3), and straight chain fatty acids (4:0, 6:0, 8:0, 10:0, 12:0, 14:0, and 18:0), which are saturated and consist of carbon chains ranging from four to eighteen carbons. Although present in trace amounts, some fatty acids contribute to the unique and desirable flavor of butter and milk fat (German & Dillard, 2010). When subjected to heat, the ß-hydroxyl fatty acids C14:0 and C16:0 undergo a spontaneous reaction to form lactones, which might enhance the flavor of butter.

The fatty acid composition of milk fat undergoes variations throughout the lactation cycle of the cow. The animal's mammary gland synthesizes fatty acids that have carbon chain lengths ranging from 4 to 14. The animal synthesizes certain 16 carbon fatty acids internally, while it acquires other portions of these fatty acids from its diet. The animal's diet contains all eighteen essential fatty acids. The energy needs of the animal and the lactation stage both contribute to systematic fluctuations in the milk fat composition. Due to insufficient fatty acids for fat synthesis in the initial phases of lactation, the animal mostly relies on its bodily reserves for energy. The fatty acids required for milk fat synthesis are obtained from the diet and typically consist of 16:0, 18:0, 16:1, and 18:2 fatty acids with longer carbon chains (German & Dillard, 2010). The levels of short chain fatty acids, specifically 4:0 and 6:0, are greater during the latter stages of lactation compared to the early stages. This is due to an increased production of these fatty acids by the mammary gland in later lactation. While the changes in the composition of fatty acids do not impact the nutritional properties of milk, they may potentially influence the processing methods of products such as butter. Approximately 65% of the fatty acids present in milk fat are saturated, whereas 30% are monounsaturated and 5% are polyunsaturated. There are variations in the nutritional properties of different fatty acids.

Saturated fatty acids are associated with both heart disease and high blood cholesterol levels. Contrarily, short chain fatty acids have four to eight carbons and undergo a distinct metabolic process compared to long chain fatty acids, which consist of sixteen to eight carbons. Additionally, they are not believed to contribute to the development of heart disease. Conjugated linoleic acid, a trans fatty acid included in milkfat, has several health advantages for individuals. The concerns regarding milk and human health are discussed in the section by German and Dillard (2010).

The triglyceride molecule exhibits a distinctive configuration of fatty acids. Typically, longer fatty acids are found in the middle and top carbon positions of the triglyceride molecule, while the majority of short chain fatty acids are positioned at the bottom. The composition of fatty acids in the triglyceride backbone affects the flavor, texture, and nutritional quality of milk fat.

2.4.2. Physical Properties of Milk Fat

Milk fat undergoes melting over a broad temperature spectrum, approximately spanning from –40°F (–40°C) to 104°F (40°C). The disparity in the hardness of butter at ambient temperature compared to its hardness when refrigerated exemplifies this phenomenon most effectively. The spread ability of butter increases with higher temperatures due to its composition, since it is around 50% solid when stored in the refrigerator and only 20% solid at room temperature. The spatial configuration of the fatty acids on the triglyceride molecule and their individual melting points are responsible for determining the melting properties of milk (German & Dillard, 2010).

The triglycerides included in milk fat possess a spherical structure. The globules in milk's serum phase are stabilized by a protein and phospholipid membrane that surrounds them. The size range of natural globules varies from less than 1 μm to over 10 μm (Gallier et al., 2020). The larger globules are capable of floating due to the uneven distribution of their sizes, a phenomenon referred to as creaming. This results in the formation of a distinct "cream line" at the top of the container. In order to minimize the separation of cream and provide a uniform dispersion of globules in the liquid phase, milk undergoes homogenization to reduce the size of larger globules to less than 1 μm.

2.4.3. Deterioration of Milk Fat

Milk fat can be degraded by oxidation, exposure to light, and enzymatic activity. Distinct mechanisms propel the advancement of each of these processes. Lipolysis refers to the enzymatic breakdown of fat by lipases. Milk lipases can originate from different sources, such as airborne bacterial contamination, intentionally introduced bacteria for fermentation, and somatic cells present in milk. Lipases remove fatty acids from the glycerol backbone of triglycerides. Lipase's enzymatic action typically imparts an undesirable acidic flavor to milk (Gallier et al., 2020).

Pasteurization prolongs the shelf life of milk and deactivates lipases. However, several types of cheese, including provolone and blue cheese, necessitate a little amount of lipolysis in order to develop their unique flavor. Milk can rapidly undergo photodegradation, leading to the development of a distinct off-flavor.

The main factor responsible for this unpleasant taste is the degradation of proteins. To decelerate this process, milk can be stored in containers that are not transparent. Another conventional chemical oxidation mechanism that might degrade milk fat is the oxygen's assault on the double bonds present in the fatty acids. The unsaturated phospholipids in milk undergo oxidation, resulting in the development of unpleasant tastes that some have described as metallic, paint-like, or fishy.

2.4.4. The Impact of Heat Treatments on Milk Fat

Milk fat exhibits a broad melting range and undergoes complete melting at approximately 104°F (40°C). HTST pasteurization procedures do not impact the nutritional value or usefulness of milk fat. Heightened heat treatments may stimulate oxidation processes, leading to the generation of undesirable tastes and the deterioration of fat. High heat treatments, such as ultra-high temperature (UHT) pasteurization, can lead to the degradation of milk fat globule membrane proteins. This process destabilizes the globules and leads them to coagulate (Gallier et al., 2020).

2.5. VITAMINS AND MINERALS IN MILK

Milk contains vitamin B12, riboflavin, and thiamin. Niacin, pantothenic acid, vitamin B6, vitamin C, and folate are present in milk in small quantities, however it is not considered a significant dietary source of these vitamins. Milk contains the fat-soluble vitamins A, D, E, and K (Gaucheron, 2005) (Table 2.2).

Table 2.1. Concentration of Vitamin and Minerals in Goat and Cow Milk

Vitamin	*Goat milk*	*Cow milk*	*Minerals (mg)*	*Goat*	*Cow*
Retinol (IU)	185	126	Calcium	134	119
Calciferol (IU)	2.3	2.0	Phosphorous	121	93
Vitamin B_1 (mg)	0.048	0.038	Magnesium	16	13
Vitamin B_2 (mg)	0.138	0.162	Potassium	204	151
Vitamin B_3 (mg)	0.277	0.084	Sodium	41	58
Vitamin B_5 (mg)	0.31	0.32	Iodine	0.022	0.21
Pyridoxine (mg)	0.046	0.042	Sulphur	28	32
Ascorbic acid (mg)	1.29	0.94	Iron	0.05	0.05
Vitamin $B_{9\ (ug)}$	1.00	5.00	Copper	0.05	0.06
Vitamin $B_{7\ (ug)}$	1.5	2.0	Manganese	0.032	0.02
Cobalamin (ug)	0.065	0.357	Zinc	0.30	0.38

Source: Chauhan et al. (2021); Park (2012).

2.5.1. Milk's Vitamin Content

Milk is a highly nutritious beverage that serves as an excellent rich source of vital vitamins. It includes Vitamin D, crucial for preserving robust skeletal structure and dental health, as well as enhancing immune system activity and promoting cellular proliferation. Furthermore, milk is an excellent reservoir of Vitamin A, an essential nutrient for the maintenance of vision, immune system, and skin appearance. Furthermore, milk supplies Vitamin B12, a vital component for the production of red blood cells and the functioning of the nervous system. Furthermore, it includes Vitamin B2 and B3, which are involved in the generation of energy and metabolic processes.

The fortification of milk with Vitamin D is a common practice aimed at mitigating inadequacies, particularly among vulnerable groups including children, pregnant women, and elderly persons. Nevertheless, it is crucial to acknowledge that the vitamin content varies among different types of milk. Whole milk offers a higher concentration of fat-soluble vitamins such as Vitamin A and D in comparison to skim milk, but it also has a greater amount of saturated fat.

Integrating milk into your diet can assist you in fulfilling your daily vitamin needs. Regardless of your preference for whole milk or skim milk, it is crucial to take both in moderation because of their high fat content. Undoubtedly, milk i□a highly beneficial re□ervoir of vital vitamin□that □ignificantly enhance general well-being.

2.5.2. Mineral Composition of Milk

Minerals have a crucial role in bone production, maintaining water balance, enzyme activity, and oxygen transport in the body. Milk contains significant quantities of calcium, magnesium, phosphorus, potassium, selenium, and zinc.

Milk contains many minerals, such as calcium phosphate, which are present a□□alt□ Milk contain□around 67% of calcium, 35% of magne□ium, and 44% of phosphate in the form of ions that are enclosed within the casein micelle. The remaining quantities are soluble in the serum phase. The presence of protein phosphate does not impact the nutritional availability of calcium or phosphate ions (Gaucheron, 2005). Milk is not considered a substantial dietary source of these minerals, although it does contain small amounts of copper, iron, manganese, and sodium.

- Vitamin D, in conjunction with calcium, aids in the preservation of robust bones and reduces the occurrence of stress fractures. Furthermore, it is essential for the upkeep of the immune system, as well as for the proper functioning of muscles and neurons, contributing to overall well-being.
- Vitamin B12 supports cardiovascular health, facilitates the production of red blood cells, and maintains the proper functioning of the central nervous system.
- Vitamin A i□a beneficial compound for □upporting many type□of physical activity as it functions as an antioxidant and aids in the maintenance of a robust immune system.
- Riboflavin is a compound. This vitamin, known as a metabolic regulator, plays a crucial role in modulating metabolism and is e□□ential for the efficient conver□ion of food into energy.
- Phosphorus, together with calcium and vitamin D, contributes to the maintenance of bone health. Phosphorus is an essential component for our bodies to produce energy.
- Niacin, a member of the B vitamin complex present in milk, plays a crucial role in regulating metabolism and facilitating the conversion of food into energy.
- Zinc is essential for immune system function and plays a crucial role in promoting normal growth and development. Iodine is crucial for protein synthesis, metabolism, and cognitive development in children. Milk is a highly effective method to replenish the body after

an intense workout because, unexpectedly, this valuable ingredient is depleted through perspiration.

- Selenium is crucial for DNA synthesis, thyroid hormone metabolism, and reproduction. This nutrient is highly labor-intensive, enabling you to increase production.
- Pantothenic acid, frequently known as vitamin B5,is essential compound contributes to recovery, metabolic control, and the transformation of food into energy.

2.5.3. Light and Heat Impact Milk's Nutrient Content

The modest heat treatment used in typical high temperature short time (HTST) pasteurization of fluid milk does not have a substantial influence on the vitamin content. Some water-soluble vitamins are depleted when subjected to the elevated temperatures used in ultra-high temperature (UHT) pasteurization, a process that increases the longevity of these products. Thiamin has decreased from 0.45 to 0.42 mg/L, vitamin B12 has been reduced from 3.0 to 2.7 μg/L, and vitamin C has decreased from 2.0 to 1.8 mg/L (Gaucheron, 2005).

Due to its thermal stability, riboflavin remains unaffected by high temperatures. The casein micelle will undergo the influx and efflux of calcium phosphate. Due to fluctuations in temperature. It is feasible to reverse this process under ambient conditions. This does not impact the nutritional properties of the milk minerals. At exceedingly high temperatures, the calcium phosphate can undergo precipitation from the solution, resulting in an irreversible alteration of the casein micelle's structure. Exposure to light causes a reduction in the levels of riboflavin and vitamin A in milk. In order to maximize the preservation of vitamins, it is recommended to store milk in containers that have the ability to block light, such as opaque plastic or paperboard (Gaucheron, 2005).

2.6. MILK ENZYMES

Specific enzymes are associated with different components in milk. The whey protein fraction includes catalase and ribonuclease, while proteinases are associated with casein micelles, and xanthine oxidase makes up a significant portion of the fat globule membrane (Fox & Kelly, 2006).

Enzymes are proteins that possess biological activities. Milk enzymes can originate from different sources, such as airborne bacterial contamination, intentionally introduced bacteria for fermentation, and somatic cells present in milk.

Milk has a diverse range of biological components, such as growth hormones, enzymes, functional peptides, oligosaccharides, and immunoglobulins, in addition to its high content of proteins, lipids, carbohydrates, minerals, and vitamins. Mammalian milk contains around sixty-three indigenous enzymes. Enzymes including lysozyme, catalase, superoxide dismutase, lactoperoxidase, xanthine oxidase, ribonuclease, and others play a crucial role in maintaining the stability of milk and protecting individuals from harmful pathogens due to their antioxidant and antibacterial capabilities (Fox & Kelly, 2006).

Each enzyme have a specific site on its target molecule where it exhibits optimal functionality, which is determined by factors such as temperature and pH. The functions of several enzymes present in milk remain poorly comprehended. The enzymes present in the human stomach and small intestine are principally responsible for the digestion of milk in humans, while the enzymes contained in milk do not have a substantial function in this process. Lipase enzymes catalyze the hydrolysis of fats. Lipoprotein lipase is the primary lipase present in milk. It is linked to the micelle of casein. Agitation during processing can result in the lipase enzyme coming into contact with the milk fat, potentially causing the breakdown of fat and resulting in unpleasant flavors. Pasteurization deactivates the lipase in milk, hence prolonging its storage duration (Fox & Kelly, 2006). Proteins are broken down by enzymes known as proteases. Plasmin is the primary enzyme responsible for breaking down proteins in milk. Heat denatures certain proteases while leaving others unaffected. Protein degradation can contribute to the desirable texture of cheese throughout the ripening process, but it can also lead to undesirable bitter off-flavors (Fox & Kelly, 2006). The cheese literature has a plethora of information regarding the involvement of proteases in cheese making. Alkaline phosphatase, a heat-sensitive enzyme found in milk, is used for pasteurization. Alkaline phosphatase is deactivated in milk by proper pasteurization. Lactoperoxidase is one of the enzymes in milk that is highly resistant to heat. The addition of hydrogen peroxide and thiocyanate to lactoperoxidase results in antibacterial properties. There is a hypothesis that the presence of lactoperoxidase in raw milk may hinder the growth of pathogens, which are germs responsible for causing diseases. However, fresh milk does not contain hydrogen peroxide and thiocyanate. Therefore, in order to give milk antibacterial qualities, these compounds would need to be added. Lysozyme, another enzyme possessing antibacterial characteristics, is present in minimal quantities in milk.

Milk has a variety of native enzymes, each serving a distinct purpose and exhibiting different levels of stability during processing. These enzymes also have varying effects on dairy products and play a crucial role in ensuring consumer safety, such as through the presence of antimicrobial enzymes. Certain enzymes, such as lactoperoxidase, possess advantageous properties, while others, like alkaline phosphatase, are employed as indicators in the processing. Plasmin and lipoprotein lipase are enzymes that can have either positive or negative impacts on the quality of dairy products, depending on the specific product.

Milk, a nourishing substance, can be transformed into many products such as cheese, kefir, cream, yogurt, butter, and yogurt by utilizing enzymes. When incorporated into animal feed, they are utilized to enhance milk production. In addition, they contribute to the preservation of milk, serve as a constituent of active packaging that prolongs the shelf life of milk, enhance the taste and consistency of milk and its derivatives, and assist in the production of lactose-free products that offer various health benefits. Enzymes are utilized to reduce the BOD (Biochemical Oxygen Demand) and COD (Chemical Oxygen Demand) of dairy effluents, serving the purpose of environmental preservation and extracting valuable products from dairy waste (Fox & Kelly, 2006).

Milk, being a perishable item, rapidly degrades during storage, primarily due to microbial activity. Moreover, numerous enzymes have been linked to various biochemical changes in both raw and processed milk, leading to reduced stability and a shorter duration of storage.

Due to its liquid state, milk is more vulnerable to enzymatic activity compared to other processed dairy products. Exogenous enzymes are utilized to induce specific modifications in the product. Both endogenous and exogenous enzymes have the ability to modify the primary constituents of milk, such as lactose, lipids, and proteins. Several indigenous milk enzymes possess significant technological uses.

Two important enzymes, sulfhydryl oxidase and superoxide dismutase, play a critical role in preserving the quality of milk. The presence of lipase, proteinase, acid phosphatase, and xanthine oxidoreductase in milk is worrisome due to their negative impact on its quality. γ-glutamyltransferase, lactoperoxidase, and alkaline phosphatase are employed as markers to determine the pasteurization status of milk. After a mastitis infection, the levels of several enzymes increase, such as catalase, N-acetyl-β-d-glucosaminidase, and acid phosphatase (Fox & Kelly, 2006). Both lactoperoxidase and lysozyme exhibit antibacterial activity. Specific enzymes have the ability to impart a desired flavor to cheese throughout the ripening process.

Enzymes have the potential to induce undesirable alterations in milk and dairy products. For instance, lipase enzymes have the ability to induce hydrolytic rancidity in milk and dairy products. Plasmin can induce bitterness and/or promote age gelation in milk that has undergone ultra-high temperature (UHT) processing. Lactase enzyme can provide a bitter taste in cream.

2.7. ADDITIONAL MINOR COMPONENTS FOUND IN MILK

2.7.1. Organic Acids

Organic acids such as citric, lactic, and acetic acids serve as examples. Citric acid is the primary organic acid found in milk. As a result of bacterial action, it rapidly disappears during storage. Lactose undergoes a process of decomposition, resulting in the formation of acetic and lactic acids.

Orotic acid is exclusively found in milk and serves as an intermediary substance in the synthesis of pyrimidine nucleotides. Orotic acid, total creatinine, and uric acid are suitable indicators for determining the milk content in a food product. The milks of different species exhibit varied concentrations of uric acid, creatinine, and orotic acid. Buffalo milk contains a lower amount of orotic acid (19.6 μg/100 ml) compared to cow milk (52.6 μg/100 ml). However, buffalo milk has a higher concentration of creatinine. The level of uric acid remains constant at 0.26 mg/100 ml, however the level of creatinine is elevated at 246 mg/100 ml compared to 167 mg/100 ml (Wright & Marangoni, 2003).

Acidity arises in unpasteurized milk produced under typical conditions. It is widely recognized that milk with a high acid content does not deteriorate. As a result, the other components of milk are preserved naturally when milk undergoes natural acidification. Fresh cow's milk is an excellent fermentation substrate for most bacteria due to its pH of 6.7, lactose content of 5%, protein content of 3.3%, and water activity close to 1.0 (Wright & Marangoni, 2003). Microbial growth can occur in milk from different species, even though their protein and lactose concentrations may vary. A multitude of bacterial strains can be detected in milk samples obtained from healthy, typical mammary glands.

Bacterial activity in milk leads to the production of acids. Lactose undergoes anaerobic breakdown, resulting in the formation of lactic acid and other organic acids, hence causing acid production. Multiple instances of sugar fermentations in milk have been documented, with variations depending on the specific microorganism involved and the resulting end products. For example, the presence of solely lactobacilli and streptococci leads to the production of lactic acid as the end product. The primary fermentation that occurs in milk is the conversion of lactose into lactic acid, which plays a crucial role in other biological processes. Due to their fermentative and saccharolytic nature, lactic acid bacteria are well-suited to flourish in milk. Typically, they will surpass other microorganisms in the competition for lactose, leading to acidification and the creation of an unfavorable environment for their rivals. Therefore, when appropriately processed, cultured dairy products contain carbon dioxide, acetic acid, propionic acid, and lactic acid (Wright & Marangoni, 2003). Fermentation can also be employed to produce some fermented milk products, in addition to ethanol production from whey. Additionally, the process of coliform fermentation can occur when ethanol, hydrogen, carbon dioxide, lactic acid, and acetic acid are synthesized. This type of fermentation serves as an example of unwanted spoilage caused by fermentation. An elevated level of coliform bacteria in milk indicates insufficient hygiene practices. Coliform fermentation not only disrupts lactic acid fermentation but also spoils cheese. The composition and diversity of bacteria will dictate the type of fermentation that ensues carbon dioxide, acetic acid, propionic acid, and lactic acid.

Several fermented milk products, such as buttermilk, kefir, and kumis, are meals that are rich in nutrients and may have potential health advantages. The technical decisions made during the process can modify the organic acids, which play a vital role in the fermentation of milk and cheese. The organic acid content of several goat cheese kinds was analyzed and recorded based on the methods used for processing. For example, researchers found differences in the concentrations of tartaric, formic, and uric acid in goat cheese that was fresh and goat cheese that had been frozen and then thawed. Upon freezing the food, the concentration of tartaric acid decreased, while the concentrations of formic and uric acids increased. The organic acid composition of the various forms of plain soft differs significantly.

There is a variation in the acidity level of simple soft goat cheese, which leads to a range of different effects.

2.7.2. Growth-Related Factors

Insulin-like growth factor I (IGF-I), transforming growth factor (TGF)-α2, fibroblast growth factor (FGF2), and specific members of the epidermal growth factor (EGF) family are the primary growth factors present in cow milk and colostrum (Wright & Marangoni, 2003). Table 6 provides a summary of the current experimental data regarding the concentration of certain development factors in cow's milk. Except for betacellulin (BTC), colostrum and milk usually contain higher quantities of this substance.

EGF and BTC, both belonging to the EGF family, have been found in milk products in significant quantities, capable of eliciting physiological effects. They enhance the growth of epidermal, epithelial, and embryonic cells; suppress the production of stomach acid, and facilitate the healing of wounds and the breakdown of bone tissue.

The TGF-α family comprises a collection of versatile growth and differentiation factors that have an impact on nearly all types of cells. Their behaviors differ according on the specific cell type, stage of cell division, and surrounding conditions. Consequently, they play a vital role in the growth of bone and cartilage, healing of tissues, formation of embryos, and modulation of the immune system. TGF-2, which is the predominant form of the TGF family found in milk products, potentially has a role in neonatal mucosal immunity or the development of the gut epithelium. However, its exact physiological function remain□unknown. TGF-κ□function a□growth □uppre□□or□for □pecific cells, including lymphocytes and epithelial cells, while promoting the growth of other cells, particularly in connective tissue (Wright & Marangoni, 2003).

IGF-I and IGF-II increase the proliferation of many different types of cells. IGF-I stimulates the growth, specialization, and maturation of cells. IGF-I further stimulates glycogen production and enhances glucose absorption. Administration of IGFs to humans leads to hypoglycemia, enhanced renal functioning, reduced cholesterol and potassium levels, and improved nitrogen balance.

Fibroblast growth factor (FGF2) exerts diverse impacts on various cell types. Reports indicate that it stimulates the migration, proliferation, and differentiation of fibrobla□t□, endothelial cell□, and epithelial cell□ Furthermore, thi□growth factor facilitates the growth and development of tissues, the generation of new blood vessels (angiogenesis), the production of blood cells (hematopoiesis), and the □ynthe□i□of collagen and fibronectin (Wright & Marangoni, 2003). The milk fat globule membrane's heparin sulfate proteoglycan has been suggested as the binding site for FGF2 in milk.

CHAPTER 3

MILK PROCESSING

Contents

Milk is considered one of the most nutrient-dense foods consumed worldwide. In order to enhance the nutritional content, the milk undergoes processing. The milk processing procedure encompasses several stages, including the collection of milk from the farm, its storage in tanks, and further processes of separation, pasteurization, and homogenization, all aimed at improving its quality (Tamime, 2009). The milk processing equipment typically comprises milk tanks for storing different types of milk, such as raw, skimmed, or cream. It also includes pasteurizers that eliminate bacteria and reduce enzymatic activity in the milk, separators that prevent contamination after pasteurization to ensure high quality, and homogenizers that improve the flavor, texture, and other sensory characteristics of the milk. High-quality milk is the final outcome of all these processes.

The milk processing procedure consists of the following steps:

- Milking on large farms is performed using automated methods twice daily. At this point, the temperature and flavor of the milk are assessed.
- The milk is then transferred to large refrigerated tanks and transported to the processing facilities by huge trucks.
- The milk is separated and clarified using separators to separate it from bacteria and useless residues.
- Following the process of separation and clarification, milk is enriched with essential vitamins and minerals.
- The milk undergoes pasteurization to eliminate harmful microorganisms by subjecting it to heat.
- Homogenization is the process by which milk fat is extracted.
- Subsequently, milk is hermetically enclosed in plastic covers or cartons. The expiration date is clearly marked, and the packs are distributed to multiple states.

Methods such as fermentation or cooling have the greatest impact on the quality of raw milk and can extend its shelf life by several days (Tamime, 2009). Pasteurization is a thermal process that extends the milk's storage time by reducing the number of hazardous microorganisms to a level that does not constitute a significant threat to human health. Further processing of milk can yield valuable, concentrated, portable, and durable dairy products such as ghee, butter, and cheese.

Small-scale dairy farmers can increase their profits by engaging in milk processing rather than selling raw milk, and processing also grants them greater opportunities to reach local and urban markets. Milk processing can help effectively regulate the fluctuations in milk supply that occur over different seasons. The establishment of employment opportunities in sectors such as milk collecting, transportation, processing, and marketing, which are not directly related to farming, can have a positive impact on entire communities. This is particularly evident in the conversion of raw milk into processed milk and its associated products (Tamime, 2009).

3.1. PASTEURIZATION

Pasteurization is a mild thermal process that entails heating both packaged and unpackaged liquid meals to a temperature below 100°C. The objective of the heat treatment and chilling technique is to inhibit the occurrence of a phase transition in the product. The acidity of the food impacts both the duration of its shelf life and the temperature and time limits of the heat treatment. The criteria also take into account heat-sensitive sensory qualities and diet.

Pasteurization refers to the procedure of subjecting a liquid to heat below its boiling point with the intention of eliminating microorganisms. The process is named after Louis Pasteur, who created it and first proposed that it could improve the wine's preservation capabilities. The significance of commercial pasteurization of milk grew in the United States during the early 1900s and in Europe during the late 1880s (Nasr, 2008). Pasteurization effectively eliminates the majority of milk-borne diseases such as typhoid, scarlet fever, septic sore throat, diphtheria, and diarrhea.

Pasteurization can be carried out using either a batch or continuous technique. At present, pasteurization is conducted either in batches at a temperature of 62.8°C for a duration of 30 minutes, or constantly at a higher temperature of 71.7°C for a shorter period of 15 seconds. However, the circumstances of pasteurization differ among different milk products. There are multiple variations of pasteurization methods, such as aseptic ultra high temperature (UHT), continuous ultra-pasteurization, continuous higher heat shorter time (HHST), and continuous high-temperature short time (HTST). Normally, canned goods undergo a process of sterilization when they are exposed to a temperature of 115.6°C for a duration of 20 minutes (Nasr, 2008).

Milk is an excellent medium for the proliferation of germs, and when stored at ambient temperature, bacteria and other harmful microbes rapidly increase. As per the US Centers for Disease Control (CDC), raw milk is considered one of the most hazardous foods globally due to the high risk of hospitalizations caused by improper handling. In fact, the number of hospitalizations resulting from raw milk is roughly three times higher than any other type of food-borne disease. Pasteurization not only eradicates *Salmonella, Listeria, Yersinia, Campylobacter, Staphylococcus aureus, and Escherichia coli*, but also serves as a preventive measure against diseases such as brucellosis, diphtheria, scarlet fever, and Q-fever (Nasr, 2008).

3.1.1. Techniques for Pasteurizing Milk

- **High-Temperature Short Time (HTST) Pasteurization:** Flash pasteurization is an alternative term for this type of pasteurization. Flash pasteurization, a process that involves heating raw milk to a temperature of 71.7°C for 15 seconds, effectively eliminates *Coxiella burnetii*, the most heat-resistant pathogen found in raw milk. Working with a range of temperatures is usually advisable as achieving the exact temperature for milk is technically unattainable. To ensure safety, it is recommended to heat the milk for 15 to 20 seconds until it reaches a temperature of 72°C to 74°C (Nasr, 2008). This will ensure that the milk continually reaches the proper temperature. Continuous pasteurization systems are the most suitable uses for this method. Flash pasteurized milk has a shelf life of 16 to 21 days.
- **Low-Temperature Long-Term Pasteurization (LTLT):** This is a process of pasteurization that involves subjecting a substance to relatively low temperatures for an extended period of time. In this process, the temperatures for pasteurization are reduced to 63°C and kept at that level for a duration of 30 minutes. Extended storage duration alters the structure of milk proteins, enhancing their appropriateness for the creation of yogurt. Batch pasteurization is most successfully achieved by storing milk in a jacketed vat. There is a wide range of batch pasteurizers available in the market, suitable for both household and commercial use.
- **Ultra-High Temperature (UHT) Pasteurization:** During the entire procedure, the product remains completely shielded, without any exposure, even for a moment. The process involves rapidly heating milk or cream to a temperature of 135°C to 150°C for a very short duration, immediately freezing it, and then storing it in a completely sterile container. UHT pasteurization remains the predominant method for preserving milk, ensuring its safety and stability, despite the potential occurrence of Millard browning (Nasr, 2008).

3.1.2. Steps of the Pasteurization Process

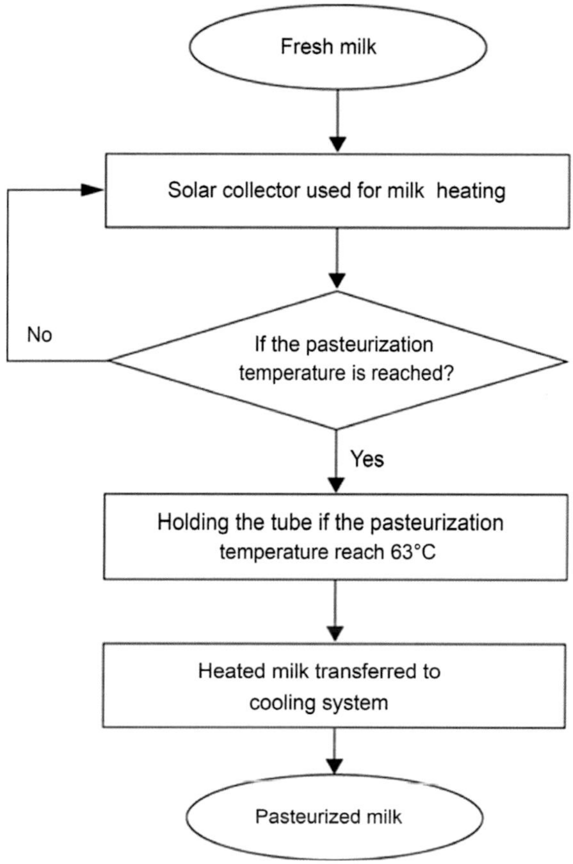

Figure 3.1. *Process flow diagram for pasteurized milk.*

Source: Ramkumar et al. (2022).

The Steps of the Pasteurization (Figure 3.1) are the explained below:

- **Cooling Milk:** Although chilling is distinct from pasteurization, it is a crucial process when dealing with large quantities of milk. When milk is expelled from the cow's udder at a temperature greater than the surrounding air, the proliferation of germs accelerates, expediting the process of spoilage. Nevertheless, the development and metabolic activities of bacteria are halted when temperatures are reduced within the range of 2°C to 5°C. Prior to the commencement of true pasteurization, this process provides a favorable beginning for ensuring quality (Elizondo-Salazar et al., 2007).

- **Pre-Heating Stage:** To facilitate the process of standardization, the chilled milk is warmed to around 40°C after being combined, which aids in the separation of butterfat (Elizondo-Salazar et al., 2007). The device utilizes regenerative heating to warm the cold milk, extracting heat from the pasteurized milk. The pasteurized milk is cooled through a countercurrent flow with chilled milk. The objective of standardization is to create a product that maintains a uniform butter-fat content.
- **Clarification:** Clarification is important in order to verify that the product is completely devoid of any extraneous substances. The milk undergoes filtration using metallic tubular filters to remove big solid particles. Milk undergoes a process of thorough cleansing to remove any dirt and sediments. This is achieved by employing a centrifugal clarifier, which is distinct from the one used for standardizing purposes. Continuous processing is facilitated by the utilization of filters, which are commonly put in parallel twins to allow for one to be cleaned while the other remains in operation (Elizondo-Salazar et al., 2007). To inhibit bacterial growth, regularly clean the filters at intervals of two to ten operating hours, depending on the level of dirt accumulation.

Proper pasteurization is crucial for the following reasons:

- To eradicate pathogenic germs that could potentially endanger public health.
- After the elimination of these microorganisms, the product becomes suitable for consumption by the general population.
- Moreover, pasteurization eliminates detrimental bacteria and enzymes that may cause the goods to decay.
- Consequently, the milk's durability is prolonged.
- Ensuring that the product has a prolonged shelf life without requiring expensive storage equipment is crucial.
- Pasteurization enhances the product's longevity by eliminating enzymes and microorganisms responsible for spoilage.

3.2 HOMOGENIZATION PROCESS

Homogenization refers to the process of evenly distributing a substance throughout a liquid, such as milk, by reducing it into tiny particles, such as the fat globules found in milk. The cream will not rise to the surface of milk that has been thoroughly homogenized. In order to disperse the fat globules, the milk is subjected to high pressure while it passes through small openings.

There are various techniques available to achieve homogenization, which is making a mixture of two liquids that do not dissolve in each other, uniform throughout. In order to do this, one of the liquids is transformed into a state where tiny particles are uniformly distributed across the other liquid. The process of homogenization involves the equal distribution of milk fat globules throughout the milk, resulting in smaller globules compared to the original size. This is a widely seen phenomenon (Samoichuk et al., 2020).

Homogenization is the process of combining two liquids that are unable to mix together in any proportion, resulting in the formation of an emulsion, which is a mixture of two or more liquids that are usually unable to mix. Primary homogenization refers to the process of decreasing the size of droplets in an already existing emulsion in order to create a new emulsion. On the other hand, secondary homogenization entails creating a new emulsion by reducing the size of droplets in an existing emulsion (Samoichuk et al., 2020) (Figure 3.2).

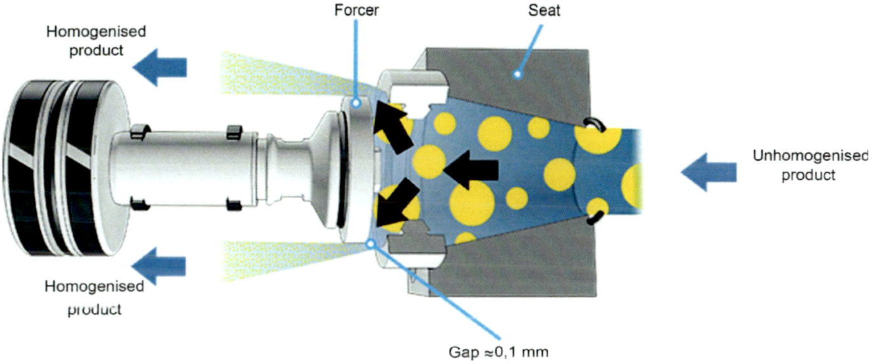

Figure 3.2. *Milk is pressurized, causing fat globules to separate during homogenization.*

Source: Dairy Processing Handbook (2019)

Homogenizing cream and other food items, such as peanut butter, can be done to achieve a stable emulsion, where the fats and oils remain uniformly mixed with other ingredients without separating. Some pharmaceutical and cosmetic products are manufactured using a comparable process.

Homogenizers are specialized pumps that are designed to handle heavy-duty tasks and operate under high pressure. They are equipped with a distinctive discharge valve. Homogenizers in the milk business are designed to disintegrate fat globules that are typically around 18 micrometers in diameter. It is worth noting that a micrometer is equivalent to one-millionth of a meter (Samoichuk et al., 2020). The application of high pressure is employed to propel heated milk, which contains fat in its liquid form, through the valve, so assuring even distribution of the fat within the milk. The advantages of homogenizing milk include a brighter appearance, enhanced taste, uniform thickness, greater ability to whiten coffee, and less curd tension for easier digestion.

Homogenization enables the enhancement of body and texture in ice cream as well as other goods such as cream cheese, half-and-half, and evaporated milk.

The process of homogenizing milk entails the amalgamation of substantial quantities of harvested milk and subjecting it to intense pressure while it is pushed through minuscule apertures. Homogenizing milk is crucial in the dairy food industry to avoid inconsistencies in fat content and flavor.

Cola products and other soft beverages exemplify the process of homogenization. To prevent the separation of constituents during transit or storage, the reactant combination undergoes extreme homogenization at pressures of up to 35,000 psi (Samoichuk et al., 2020).

3.2.1. Benefits of Homogenized Whole Milk

By increasing the viscosity of milk, it prevents the formation of cream, resulting in a more substantial appearance when added to tea or coffee. The presence of fat globules makes it unnecessary to mix the milk before serving, as they are resistant to rising.

When homogenized milk coagulates, it develops a more concentrated and dense texture, as well as a more vibrant appearance. This is partly due to the presence of smaller fat globules, which enhance the texture and flavor. Additionally, the reduced curd tension also contributes to these improvements. According to Olorunnisomo et al. (2014), it is feasible to suggest homogenized milk as a suitable option for babies.

Reduces the likelihood of fat separation during the process of creating ice cream and evaporated milk, resulting in a more uniform and velvety texture of the final product.

Reconstituted milk can be produced by combining butter or butter oil with skim milk using a homogenizer. Consequently, the milk is less prone to the development of oxidized flavors. Homogenization extends the freshness of milk. Homogenized milk contains fat cells that are consistently sized, allowing it to maintain freshness for a longer period compared to milk that is obtained immediately after milking cows. To achieve milk homogenization, it is necessary to eliminate the cream that rapidly forms a layer on the surface of the milk and diminishes its freshness. Homogenized milk is easily digested. The majority of individuals who ingest non-homogenized milk have gastrointestinal discomfort afterwards.

Homogenization of milk enhance□both it□color and flavor. Due to its perceived purity, white milk is generally preferred and chosen for purchase and consumption over creamed milk by the majority of people. According to Olorunnisomo et al. (2014), homogenized milk has a whiter hue. In addition, the process of homogenizing milk enhances its fat content and imparts a smooth and rich flavor, both of which are desirable attributes for customers.

3.2.2. Drawbacks of Homogenized Milk

Consuming homogenized milk can have negative effects on your well-being. The particles in homogenized milk are of a smaller size compared to those in non-homogenized milk. As a result, tiny particles are immediately taken into the bloodstream during digestion, posing a risk to your health. In addition, there is a correlation between the consumption of homogenized milk and the development of heart disease and cancer (Olorunnisomo et al., 2014). Homogenized milk has reduced nutritional value. Vitamins such as vitamin D and A are also fragmented into microscopic particles during homogenization due to the breakdown of milk fat into tiny particles. The value of certain nutrients in milk is lowered due to their smaller particle sizes compared to others. homogenizing milk can be advantageous as well as disadvantageous. The health problems arise most when you consume the homogenized milk in excess.

3.3. STANDARDIZATION

Milk standardization involves adjusting the levels of milk fat and solids to adhere to predetermined criteria. This technique entails modifying the constituents of milk, such as fat, protein, non-fat solids, and total solids. Standardized milk is typically available for purchase in supermarkets, whereas non-standardized milk is directly delivered to consumers from the farm. The standardization process commences by segregating cream and milk, and subsequently recombining the two components post-separation. The ideal fat content for various types of milk or cream products is achieved by restoring only the required quantity of fat. Standardization may also entail modifying the protein content as needed for the manufacturing of different dairy products. Standardization is employed in several areas of the dairy industry to quantify the milk components found in different milk-based products, including cheese, yogurt, ice cream mix, cream, whey, processed and raw milk, and homogenized milk. The study of these goods mostly involves quantifying constituents such as fat, protein, lactose, and solids.

Legal regulation□exi□t regarding the butterfat content in fluid milk, which is eaten by consumers. Skim milk generally contain□le□□than 0.10% butterfat, but low-fat option□□uch a□1% and 2% milk have a larger fat percentage. Whole milk has a butterfat content of 3.5%. Producers are required to do milk testing to verify that it complies with the fat specifications of the particular product they are manufacturing. California imposes new minimum limits for non-fat fluid milk solids.

Cheese producers oversee the levels of fat and protein in milk, with casein constituting a substantial proportion of the protein found in cow's milk. The casein to fat ratio is a crucial metric for assessing cheese yield and quality in the manufacturing process. Cheese producers assess the protein and fat content in every batch of cheese to guarantee uniformity and excellence.

Manufacturers must have knowledge of the fat and total solids composition for each batch in the production of yogurt and ice cream. The butterfat content in ice cream often falls within the range of 10% to 16%, whereas the cream utilized in these products generally has 40% butterfat.

Standardization can be achieved using three distinct approaches: batch, continuous, and automation. Small and medium-sized dairies frequently utilize batch standardization. The process entails measuring and storing whole milk to ascertain its fat content, separating skim milk and cream from the remaining milk, and subsequently quantifying and combining the necessary amount of skim milk or cream with the bulk milk. A secondary examination is performed to verify that the fat content falls within the acceptable range, and any necessary adjustments are implemented accordingly. Batch standardization necessitates the allocation of time for the processes of agitation, testing, and final blending.

Continuous standardization entails the use of an inline milk fat sampler and a testing instrument to periodically measure and exhibit the fat content. The operator manipulates the settings and valve displacement in order to mix the cream with the skim milk line and attain the targeted fat content.

Automated standardization is a recent advancement that substitutes the operator with a microprocessor unit linked to a sensor system. The microprocessor unit modulates the control valve by utilizing inputs from the sensor system in order to regulate the amount of cream that is incorporated into the skim milk. This automated approach guarantees greater accuracy and consistency while reducing the need for manual work and saving time.

To summarize, the process of standardizing milk entails altering the levels of milk fat and solids to adhere to particular criteria. This procedure is crucial in the manufacturing of different dairy products and is accomplished by separating, quantifying, and combining the required components. The many segments of the dairy business employ standardization practices to guarantee uniformity and excellence in their products.

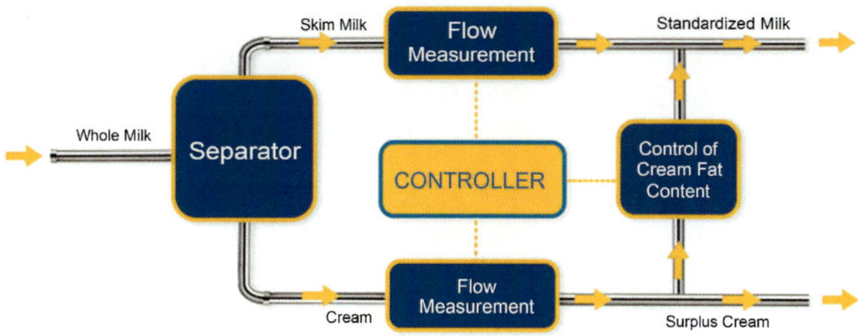

Figure 3.3. *Graphic shows closed-loop feedback control system for automatic milk fat standardization.*

Source: Neologic Engineers, 2024.

3.4. FORTIFICATION

Fortification refers to the act of supplementing cow's milk with additional vitamins and minerals that are not naturally present in sufficient quantities. The most commonly supplemented vitamins in milk are usually vitamin D and vitamin A, however additional nutrients such as zinc, iron, and folic acid may also be included. Prior to the processes of homogenization and pasteurization, heat-resistant forms of vitamin A and vitamin D3 are included (Ocak & Rajendram, 2013) (Figure 3.4).

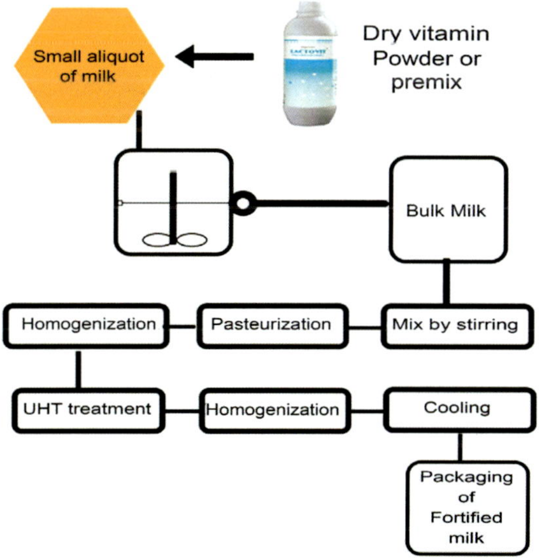

Figure 3.4. *Liquid milk fortification.*

Source: Pristine Premixes, 2024.

The utilization of fortified and unfortified milk is indistinguishable. It is appropriate for both culinary and beverage purposes. Fortifying with B vitamins is not a frequent practice, despite the fact that these nutrients are heat-sensitive and are typically added after the heating process.

Fortified milk is milk that has been enriched with additional nutrients. While not legally required, milk in the United States is commonly enriched with vitamins A and D. Fortified milk is a valuable source of vitamins A and D.

In order to ensure that plant-based milk can adequately substitute regular cow's milk, it is necessary to fortify it. Plant-based milks, although intended to resemble ordinary milk, do not possess the same nutritional composition as cow's milk (Ocak & Rajendram, 2013). There are more affordable alternatives available for individuals who are unable to buy or eat cow's milk, although these alternatives may not provide all the essential nutrients, especially calcium and protein. In order to attain nutritional equivalence, plant-based milks must be fortified with important components that are crucial for growth and development. For example, milk alternatives derived from grains such as rice and oats could lack sufficient amounts of calcium and protein. Fortification aids in meeting clients' nutritional requirements by minimizing the necessity for additional supplements. As an illustration, soy milk contains around half the calcium content of cow's milk, while having a similar protein content. In order to address this scarcity, fortifying agents such as tri-calcium phosphate and calcium carbonate are employed. Nutrient depletion can occur by processing methods such as blanching, flaking, and ultra-high temperature treatment, especially for vitamins and minerals. During the process of fortification, nutrients are restored to compensate for the aforementioned losses (Ocak & Rajendram, 2013). Plant-based beverages can be included in the diet of children over the age of two who cannot consume cow's milk for medical reasons. However, it is important to ensure that these fortified plant-based beverages contain at least 6 g of protein per 250 ml to maintain a balanced diet.

3.4.1. Benefits of Milk Fortification

Fortified milk offers several benefits compared to unfortified milk.

- **Addresses Nutritional Deficiencies:** The primary objective of fortification, which entails supplementing a meal with nutrients that it lacks, and enrichment, which involves restoring minerals that are lost during processing, was to prevent nutrient-deficiency disorders such as rickets, a condition characterized by weakened bones due to a deficiency in vitamin D. In affluent countries, the occurrence of deficiency diseases has nearly vanished due to the process of enhancing and fortifying milk and bread. In addition, less severe but still potentially dangerous deficiencies in micronutrients can be remedied through fortification.

- **Promotes Optimal Development in Children:** Iron deficiency anemia is a common illness in children, especially in developing countries, and can be prevented by consuming milk that has been fortified with iron. Milk in these places is often fortified with iron and other essential minerals such as zinc and B vitamins. An analysis of trials including over 5,000 children (Ocak & Rajendram, 2013) found that supplementing milk and grain diets with iron, zinc, and vitamin A resulted in a nearly 50% reduction in the prevalence of anemia in children under five years old.
- **Improves Skeletal Strength and Density:** Fortified milk may enhance bone health by providing additional nutrients. The utilization of fortified dairy products such as milk is associated with an increase in bone mineral density, resulting in stronger and thicker bones. Milk is rich in calcium and phosphorus, which are essential nutrients for bone formation. Thus, even without fortification, milk can contribute to bone health by providing the necessary components to create and enhance bone structure. However, milk that has been enriched with vitamin D is especially beneficial for preserving strong bones as it enhances the body's calcium absorption. Osteoporosis is a medical illness characterized by the weakening and brittleness of bones. It can be prevented by following a diet that is rich in calcium (Ocak & Rajendram, 2013). An economical and convenient approach to guarantee enough calcium intake and enhance absorption of this essential mineral is to consume fortified milk.

3.5. FERMENTATION

Fermentation is the enzymatic conversion of lactose, a milk sugar, into lactic acid by the action of helpful microorganisms. This technique not only enhances the nutritious composition of milk but also imparts a tangy taste and aids in its preservation. The bacteria utilized in the process of milk fermentation, specifically *Lactobacillus delbrueckii* bulgaricus and Streptococcus salivarius thermophilus, metabolize lactose molecules and transform them into pyruvic acid. In the absence of oxygen, this pyruvic acid is further converted into lactic acid (Yousseef et al., 2016).

Yogurt and other fermented milk products are more easily digested by people who have lactose intolerance or milk protein allergies. The process of fermentation enzymatically decomposes proteins and lactose in milk, rendering it more easily digestible for persons with such sensitivities. In addition, fermented dairy products such as kefir and yogurt include a high amount of probiotics, which promote gastrointestinal health and enhance the immune system.

Fermenting milk offers numerous benefits. Firstly, it enhances the ability to digest food, especially for individuals with lactose intolerance. Lactose is broken down by the process of fermentation, resulting in the increased digestibility of milk. Furthermore, fermented milk products include augmented nutritional value as a result of the inclusion of probiotics. Probiotics have demonstrated several health advantages, including as promoting gastrointestinal well-being and enhancing the immune system. Finally, the process of fermentation extends the duration for which milk may be stored without spoiling, hence decreasing the likelihood of spoilage (Yousseef et al., 2016).

Home fermentation of milk can be accomplished by straightforward procedures. Begin by selecting high-quality pasteurized milk, such as whole milk or low-fat milk. Bring the milk to a temperature of around 180°F (82°C) in order to eliminate any potentially hazardous bacteria and enzymes that could disrupt the fermentation process. Allow the milk to reach a temperature of approximately 110°F (41°C), as this is the optimal temperature for the proliferation of beneficial bacteria. Incorporate a modest quantity of live yogurt or a yogurt starter culture into the milk to introduce the essential microorganisms required for fermentation. Transfer the milk into a sterilized, hermetically sealed container and position it in a heated location, such as a yogurt maker or an oven with the light turned on, for a duration of six to twelve hours to allow for incubation. After achieving the appropriate level of sourness, cool the fermented milk to stop the fermentation process and maintain the flavor (Yousseef et al., 2016).

Yogurt and other fermented milk products are produced by introducing milk to a lactic acid bacteria culture sometimes referred to as a "yogurt culture." The bacteria in question undergo lactose fermentation, resulting in the production of lactic acid. This lactic acid then interacts with milk protein, ultimately contributing to the unique flavor and texture characteristic of yogurt. The choice of lactic acid bacteria, such as Streptococcus thermophilus, might impact the consistency of yogurt. The protein content in milk is also a determining factor, as larger amounts of protein lead to the production of yogurt with a thicker consistency. For instance, Greek yogurt has a greater amount of protein and has a denser texture. There is an increasing need for yogurts that are high in protein and can be consumed easily, without being as thick as typical yogurts (Schmidt et al., 2016).

Fermented milk has been utilized for diverse health objectives, including the management of lactose intolerance, hypertension, hypercholesterolemia, dermatitis, antibiotic-induced diarrhea, and common respiratory infections.

Nevertheless, a significant number of these applications lack substantial scientific data to substantiate their efficacy (Schmidt et al., 2016). Dairy goods such as ice cream, cheese, yogurt, and other cultured milk products are consumed globally. These products are derived from the process of fermentation and provide a diverse range of flavors and textures (Youssef et al., 2016).

3.6. CONCENTRATION

In the dairy industry, the typical technologies employed to concentrate milk are reverse osmosis (RO) or thermal evaporation. Concentrating heat requires a significant amount of energy and leads to alterations in the flavor and color of the final product. In addition, fouling has a negative impact on reverse osmosis (RO) and reduces the maximum concentration of the product that may be attained (Balannec et al., 2002).

Membrane technology can be utilized to concentrate and purify food components, hence maintaining the original flavor compounds of the food. It has been widely used in the process of concentrating skim milk. High-quality ice cream can be made by concentrating cow milk using a nanofiltration membrane. In general, condensed milk contains a high amount of salt, which negatively affects the taste of the resulting ice cream (Balannec et al., 2002). The ice cream has a smooth and creamy texture due to the reduction in salt content of the milk condensed by the nanofiltration membrane. Furthermore, the product possesses a pronounced milk flavor as a result of its absence of heating.

Ultrafiltration is a process that allows lactose and ash to flow through, while effectively retaining almost all of the protein present in raw milk. To minimize the presence of lactose and ash, a thorough filtration process can be employed. This involves continuously adding water to the retained liquid and repeating the filtration process to get a concentrated milk protein with a protein concentration exceeding 85%. Moreover, this technique is utilized in the production of desalted and de-lactose whey powder, which is utilized in the creation of ice cream, yogurt, soft cheese, and other fermented dairy products, as well as high protein skimmed milk powder. In addition, it can be utilized alongside electrodialysis to produce whey protein concentrate, which contains 96% solid content, 50% lactose, 35% nitrogen, and 8% ash content (Balannec et al., 2002).

Ultrafiltration concentration has been widely employed in the production of cheese. Various concentration ratios are utilized to concentrate the raw milk for cheese production. This process increases the concentration of the raw milk, preserves the starter and rennet, and reduces production in subsequent stages. By doing so, the amount of casein and whey lost is reduced, resulting in an increase in cheese production.

The primary classical concentration techniques utilized in large-scale dairy manufacturing are evaporation and spray drying methods. These unit operations decrease the water content by employing elevated temperatures (Balannec et al., 2002). Excessive heat treatment that exceeds the thermal stability of milk can lead to the creation of sediment, grittiness, phase separation of milk fat, and other unpleasant sensory and physiochemical alterations.

Novel non-thermal techniques have been developed and examined in the dairy sector as prospective substitutes. These alternative procedures maintain the integrity of food colors, original volatile components, vitamins, enzymes, and proteins. They also enhance the production of high-quality dairy products and minimize the negative impacts of traditional concentration processes.

Membrane separation, freeze-drying, and freeze-concentration technologies are efficient, gratifying, and capable of substituting traditional concentration procedures in the manufacturing of superior concentrated dairy products.

3.6.1. Conventional Concentration Processes

Various food techniques, including traditional thermal methods or advanced technology, can be used to extract the liquid component and produce concentrated or dehydrated dairy products. The desired effect on the structure of the product and the required level of concentration will decide the optimal choice for the process. Considering that the concentration and drying processes, or their combination, are the most energy-intensive operations in the dairy sector, it is crucial to consider the overall costs associated with these procedures before implementing them on a large scale (Beldie & Moraru, 2021).

- **Evaporation:** The main objective of the process of evaporating milk, which has been practiced for many years, is to increase the concentration of solutes such as proteins, lipids, carbs, and minerals by eliminating water from the solution. The objective of employing elevated temperatures is to prioritize the minimization of the whole expenditure, including capital and operational costs, product wastage, energy consumption, and cleaning expenditures.

Evaporators are commonly designed to facilitate the transfer of concentrated liquid through a heat-receiving tube from the external environment. Water is evaporated from the concentrated solution by heating it to its boiling point under normal atmospheric pressure, which is 100°C at sea level and 85°C at an altitude of approximately 5000 m above sea level (Beldie & Moraru, 2021). Utilizing multiple stages of evaporators or vapor recycling to recover energy and decrease the demand for this source can significantly enhance energy efficiency due to the substantial latent heat of water evaporation.

In the dairy business, the evaporation process is commonly conducted using long-tube vertical falling film evaporators. In this process, the milk, which is heated to a temperature close to boiling, is poured continuously into the upper part of the inner surface of a stationary tube that is positioned alongside other tubes and covered by a protective layer. The thermal energy from the steam induces the milk to reach its boiling point as it traverses each tube and coalesces into a delicate film. The concentrated liquid is removed from the steam in a secondary separator subsequent to its separation at the base of the apparatus. The concentrated liquid is transferred to subsequent stages in evaporators with different levels of concentration, and the steam is utilized to provide heat to those stages (Beldie & Moraru, 2021).

- **Alternative Concentration Processes:** The dairy sector is increasingly focusing on alternative concentration strategies to mitigate the adverse effects of conventional processing technology. Non-thermal procedures effectively maintain the maximum levels of milk bioactive compounds, resulting in a concentrated dairy product that possesses exceptional quality, nutritional value, and functional benefits. Additionally, this product is pleasing to the senses and its production costs are fully compensated. The latest study findings allow enterprises to use innovative technologies in dairy manufacturing, taking into account the production flow, operating mode, and kind of dairy product (Beldie & Moraru, 2021). The reason for this is that alternative technologies generate goods with distinct characteristics for each method, which thus becomes a determining factor for the industrial sector in terms of widespread implementation.
- **Freeze Concentration:** The freeze concentration method entails the solidification of the water component, transforming it into unadulterated ice crystals, and subsequently extracting it from the concentrated solution. It is a method of concentrating liquid meals without using heat. The process relies on a low-temperature solid-liquid separation. This approach offers a fascinating alternative for producing dairy products as it effectively preserves their flavor, volatile compounds, and nutritional content. It can be utilized to concentrate or pre-concentrate components that are sensitive to heat.

To avoid simultaneous freezing of all components, the technique involves lowering and regulating the temperature of the solution below its freezing point, namely at the eutectic point. Subjecting the solution to controlled freezing at a temperature above its eutectic point enhances the transparency of ice crystals while preserving the properties of the original liquid matrix. In general, the maximum limit of freeze concentration for liquid food is often between 40 and 50% of solid content, depending on the number of soluble solids and the composition of the food matrix (Beldie & Moraru, 2021).

3.7. PACKAGING

Figure 3.5. *Creative milk packaging designs.*

Source: Anush (2019).

The selection of suitable packaging for dairy products (Figure 3.5) involves considering various factors such as product compatibility, chemical inertness, non-toxicity, impact resistance, tamper resistance, sanitation maintenance, light and odor protection, specifications (size, shape, weight), printability, and marketing appeal. Ultimately, all these considerations are weighed against cost constraints (Moss, 2012). The selection of material and technique for packaging dairy products is based on their inherent characteristics and qualities.

Because dairy goods like butter are prone to oxidation, they need packaging materials that have a high resistance to oxygen. Heat-resistant packaging materials are necessary for liquid milk that has undergone thermal treatment after packaging. Aluminum foil is a favored packaging material for various dairy products and serves as an effective barrier against flavors, light, and oxygen. According to Moss (2012), the paperboard provides the packaging with enhanced durability and firmness. Tetra Pak, a type of packaging made from plastics and laminates, is highly suitable for storing liquid milk due to its internal polyethylene layer that provides effective protection for the enclosed liquid. The polyethylene layer of the lamination will provide protection to the product by preventing moisture from entering from the outside.

Lacquered tinplate containers are the optimal choice for wholesale dealers looking to package ghee with a storage capacity ranging from 15 to 250 L/kg. Ghee placed in tinplate containers has a shelf life of approximately one year and is relatively stable. Tinplate containers are hermetically sealed to prevent any air from entering, as ghee is highly susceptible to oxidation. According to Moss (2012), retail packs with capacities ranging from 50 ml to 1 L made from multi-layer co-extruded films of LDPE/HDPE and polyethylene are more cost-effective compared to tinplate containers.

Ultra-high-temperature (UHT) processing extends the shelf life of milk by changing its very perishable nature. The liquid milk undergoes a heating procedure for a duration of 1–8 seconds at a temperature ranging from 135 to 154°C. Effective packaging is essential for UHT products as adequate processing alone cannot guarantee an extended shelf life for milk. UHT milks can be contained in plastic bottles and pouches, similar to pasteurized and sterilized milks (Moss, 2012). During the packaging process, the bottles can undergo blow-molding or pre-molding, which can be done in either an unsterilized or sterilized condition. UHT milk packages are made with many laminates, including a triple layer of high-density polyethylene, in order to prolong their shelf life. Typically, the entry of oxygen during storage is not a problem. Aseptic techniques are employed to fill and seal bottles, irrespective of their origin, in order to prevent contamination. Milk should remain fresh for a duration of six months when stored at room temperature. Packaging containers designed for perishable items with a limited shelf life (10–15 days) are commonly made from a combination of paper and white polyethylene. Additionally, plastic (HDPE) bottles are used to package powdered skim milk. In recent times, there has been a growing trend in the use of flexible plastic packaging for these types of products. The pliable sachets are frequently enclosed in paperboard cartons. Flexible materials, ranging from simple polyethylene bags to advanced multiplex laminates, have made significant advancements (Moss, 2012).

The criteria for packaging milk and milk products include maintaining storage rooms for packing materials that are free from dust and rodents, as they present an unacceptable risk of product contamination. Furthermore, it is imperative to maintain a clear separation between these locations and any zones that may potentially harbor substances capable of contaminating the product. According to Still et al. (2013), it is prohibited to directly place packaging materials on the ground.

The packaging process must be carried out swiftly and overseen by a separate team of professionals that possess specialized expertise in product packaging.

Containers holding pasteurized milk and milk products must be filled or bottled by hand, and the containers must be sealed automatically.

Reusable dairy product wrapping and packaging must be capable of undergoing thorough cleaning and disinfection before being utilized again.

Sealing can be performed promptly following the filling process in the plant where drinking milk or liquid milk-based products have completed their final heat treatment. This guarantees that the milk is protected from any detrimental influence from external factors that could affect its qualities (Still et al., 2013). The sealing mechanism should be designed to ensure that, even after the container has been opened, it remains readily visible and verifiable that it has been tampered with.

The dairy sector is modifying its packaging methods and technologies to more effectively fulfill the requirements of consumers and the food industry. The use of creative packaging solutions for dairy products offers numerous possibilities for enhancing product preservation and prolonging the duration of shelf life (Still et al., 2013). Certain cutting-edge approaches have reached a level of advancement where they can effectively disclose information regarding the freshness of a product.

CHAPTER 4

DAIRY PRODUCTS

Contents

Food items made from or containing milk are known as dairy goods or milk products. The most commonly favored dairy animals include water buffalo, nanny goats, ewes, and cows. Dairy products encompass commonly recognized food items available in supermarkets globally, such as cheese, yogurt, milk, and butter. A dairy is a manufacturing facility that produces dairy products. Globally, individuals use dairy products to varying extents. Some individuals choose to abstain from consuming dairy products due to lactose intolerance, adherence to veganism, environmental concerns, or other health-related reasons or beliefs.

Milk, the predominant component in dairy products, has served as a fundamental source of sustenance for humanity across endless generations. These comprise dairy products such as butter, ice cream, kefir, cheese, and yogurt (Tunick & Van, 2015) (Figure 4.1).

Figure 4.1. *Some common dairy products.*

Source: Solan (2019).

4.1. BUTTER

Dairy products such as butter are derived from the lipids and proteins found in milk and cream. In the United States, butter is predominantly derived from cow milk. However, it can also be produced from the milk of several other animals such as sheep, goats, buffalo, and yaks, among others (Wilbey, 2009). By churning milk or cream, the process of separating the fat from the buttermilk is achieved, allowing manufacturers and home cooks to produce butter. Occasionally, food coloring and salt are also included. Despite not containing genuine butter, nondairy "butters" such as cocoa, apple, and peanut butter possess a texture that closely resembles that of butter.

Churning is the process by which dairy cream is transformed into butter. A separator is used to divide the incoming milk into two distinct categories: skim milk, which contains 0.1% milkfat, and cream, which contains 35–40% milkfat. This process is carried out in order to obtain the cream. Buttermilk is produced by churning cream until little butter grains form (Wilbey, 2009).

Once the buttermilk has been drained, the grains are further kneaded together through churning. Typically, salt is used into butter to enhance its taste and prolong its storage duration. Approximately twenty liters of whole milk are needed to produce one kilogram of butter. An emulsion of water and oil is the end result of the churning process. Approximately 80% of the final result consists of milkfat, while the remaining 16% is comprised of water in the form of tiny droplets. The smooth and soft texture of butter is strongly linked to the process of kneading. Butter typically consists of water, milkfat, small amounts of protein, calcium, phosphorus, and several fat-soluble vitamins (A, D, and E), along with a salt content of 1.5–2.0% (Wilbey, 2009).

In order for butter to be easily spreadable and rapidly melt in the mouth, it should possess a uniform hue and a velvety texture. Sweet cream butter, derived from pasteurized cream, is the predominant kind of butter among the various available variants. Recent advances include light butter, reduced-fat butter, and spreadable dairy blends that have the ability to melt at refrigerator temperatures.

Studies indicate that consuming high-fat dairy products, such as butter, may lower the likelihood of developing heart disease, diabetes, and obesity. Nevertheless, it is important to note that butter contains a significant amount of saturated fat and calories. Despite being commonly dismissed as unhealthy, the health benefits of butter have been a subject of debate among specialists for several years (Wilbey, 2009). Butter is mostly composed of milk fat, making it high in fat content. Nevertheless, this ubiquitous food item is more intricate than its mere fat content.

4.1.1. Butter Nutritional Information

A single tablespoon of unsalted butter contains 102 calories, 12 grams of fat, and no fiber, sugar, carbohydrates, or protein. Butter is a rich source of essential nutrients such as Vitamin A, Vitamin D, Calcium, and Vitamin E.

4.1.2. Butter Varieties

There is a wide range of butter variations that serve different purposes.

- **Butter Made from Grass:** Butter derived from cows that are fed on grass offers certain health advantages to counterbalance the associated health risks. When comparing it to ordinary butter, it appears to have a higher proportion of unsaturated fats and a lower proportion of saturated fat. This is because butter marketed as "grass-fed" is derived from cows that are not fed high-grain diets but are allowed to graze. Research indicates that consuming grass-fed butter in moderate amounts may provide greater health benefits and pose fewer risks compared to consuming normal butter (Wilbey, 2009).

- **Whipped Butter:** This is butter that has been beaten or whipped to provide a lighter and fluffier texture. In order to increase the volume, one can aerate ordinary butter by beating it with nitrogen gas. It is advisable to apply whipped butter as a spread rather than incorporating it into recipes. Due to the variance in density, recipes will not yield identical results (Fearon, 2011). Whipping butter with air or other gases creates a product that has greater volume (lower density) and is easier to spread when it's cold.
- **Reduced-Fat Butter:** This particular type of butter contains around 40% less milk fat. Water and gelatin serve as substitutes for bovine milk fat. Using this type of butter as a substitute for regular butter in baking or pan-frying is not recommended due to its higher water content (Fearon, 2011).
- **Ghee or Clarified Butter:** At home, one can separate melted butter from the milk solids and water to create clarified butter, which is also referred to as ghee. Remove the frothy layer that forms on the surface of melted butter in a skillet. Subsequently, eliminate the clarified butter and dispose of any residual milk solids settled at the bottom. Ghee is a commonly utilized ingredient in Indian cuisine, known for its abundant, nutty taste. Due to its high smoke point, it is perfect for high-heat frying and sautéing.

4.1.3. Health Benefits of Butter

Butter is rich in beta-carotene, which the body converts into vitamin A. Research conducted by Fearon in 2011 provides data supporting the claim that beta-carotene has a mitigating effect on the likelihood of developing prostate and lung cancer. It could improve one's eye health. Moreover, the presence of beta-carotene in butter may contribute to a decrease in the likelihood of developing age-related macular degeneration and experiencing visual impairment. It could potentially facilitate bone fortification. Butter contains Vitamin D, a crucial nutrient for bone building and development. Furthermore, it has calcium, an essential element for maintaining robust bone structure. Furthermore, calcium is known to be effective in preventing osteoporosis, a medical condition characterized by bone weakening and fractures (Fearon, 2011).

It has the potential to enhance skin health. Butter contains Vitamin E, which plays a crucial role in maintaining healthy skin. This nutrient mitigates the harmful effects of UV sun exposure, alleviates skin irritation, and facilitates the recovery of skin injuries.

4.1.4. Possible Risks of Butter

Butter is rich in calories and fat, particularly saturated fat, which has been associated with the development of heart disease. The American Heart Association advises individuals to restrict their own consumption of saturated fat. Therefore, it is recommended to use this ingredient sparingly, particularly for individuals with heart disease or those aiming to lower their calorie intake (Fearon, 2011).

4.2. CHEESE

Cheese is a dairy product produced by curdling the casein protein in milk, resulting in a range of flavors, textures, and forms. It consists of proteins and fat derived from milk, usually obtained from cows, buffalo, goats, or sheep. To induce the coagulation of casein, milk is commonly acidified during processing and subsequently supplemented with either rennet or bacterial enzymes that have similar activity (Walther et al., 2008). Once separated from the liquid whey, the solid curds are compressed to form the final cheese product. Certain cheeses include aromatic molds distributed throughout, on the rind, or on the outer layer.

There exist more than one thousand distinct types of cheese produced in various countries. Several factors influence the styles, textures, and flavors of milk, including its origin (including the animal's nutrition), pasteurization, butterfat content, presence of germs and mold, processing method, and aging duration. Wood smoke, spices, and herbs can be utilized as flavorings. Annatto is commonly added to several cheeses to give them a yellow to scarlet hue. Garlic, cranberries, chives, and black pepper are among the several supplementary components. A cheese monger (Figure 4.2), also referred to as a specialized cheese merchant, possesses expertise in the careful curation, procurement, acceptance, preservation, and maturation of cheeses (Walther et al., 2008).

Bacteria metabolize lactose into lactic acid, resulting in the acidification of the majority of cheeses. Subsequently, rennet is introduced to complete the coagulation procedure.

Cheese is highly prized for its abundant fat, protein, calcium, and phosphorus content, as well as its portability and long-lasting quality. When comparing cheese to milk, it is evident that cheese is denser and can be stored for a longer period of time. However, it is important to note that the shelf life of different types of cheese may vary.

Figure 4.2. *Cheese monger.*

Source: culture: the word on cheese (2018).

Hard cheeses, such as Parmesan, have a greater longevity compared to soft cheeses like Brie or goat's milk cheese. Some cheeses can be sold under favorable market conditions because they have a long shelf life, especially when they are coated with a protective rind. In the 21st century, vacuum packing became widely used for block-shaped cheeses, while plastic bags holding mixtures of carbon dioxide and nitrogen were flushed with gas for distributing cheese (Walther et al., 2008).

4.2.1. Curdling

This refers to the process of separating milk into curds and whey, typically by the addition of an acid or the action of enzymes. The process of cheese production involves the separation of milk into two distinct components: liquid whey and solid curds. Typically, rennet and acidification are employed to accomplish this. Under specific circumstances, acidification can be accomplished immediately by introducing an acid, such as vinegar. Instead, starting bacteria that convert milk sugars into lactic acid are utilized more commonly. The ultimate taste of matured cheeses is predominantly influenced by the same bacteria and the enzymes they produce. The production of most cheeses typically involves the utilization of bacteria derived from the *Lactobacillus*, Streptococcus, or *Lactococcus* species (Walther et al., 2008).

Although the majority of fresh cheeses contain rennet, there are few that are curdled solely through acidity. Rennet induces the formation of a resilient and elastic gel in cheese, in contrast to the fragile curds that result from natural acid coagulation. Moreover, it allows for curdling to occur at a lower level of acidity. This is important since excessive levels of acidity might impede the growth of microorganisms that contribute to flavor (Walther et al., 2008). In general, harder, bigger, and more mature cheeses require less rennet for coagulation compared to softer, smaller, and younger cheeses. Currently, the predominant method of producing rennet for cheese manufacture involves the use of recombinant technology. However, in the past, rennet was obtained by extracting the inner mucosa of the fourth stomach chamber of young, unweaned calves that were slaughtered. The majority of the applied chymosin is maintained in the whey, with just minimal quantities being present in the cheese.

4.2.2. Management of Curd

Currently, the cheese has congealed into a highly moist gel. Once specific soft cheeses have undergone the processes of draining, salting, and wrapping, they are essentially completed. Divide the curd into little cubes for the majority of the remaining part. Consequently, water may leak from the individual curd particles.

Subsequently, many types of firm cheese undergo a heating process that raises their temperature to a range of 35 to 55°C (95 to 131°F) (Kapoor & Metzger, 2008). This process extracts a greater amount of whey from the finely cut curd. By manipulating the bacterial culture and adjusting the chemistry of the milk, the flavor of the final cheese is also altered. Cheeses that require higher temperatures during production often utilize thermophilic starter microorganisms, such as lactobacilli or streptococci.

4.2.3. Getting Ripe

Fresh cheese usually exhibits a resilient texture, a mild taste, and a savory flavor. This is particularly accurate when dealing with more challenging varieties. Cheese curds are highly appreciated for their distinctive characteristics, but, in general, cheeses are permitted to mature in meticulously controlled conditions. The process of aging, often referred to as ripening, can vary in duration from a few days to several years (Kapoor & Metzger, 2008). Enzymes and bacteria alter the texture and enhance the flavor of cheese as it matures. This alteration is principally caused by the breakdown of casein proteins and milkfat into a complex mixture of fatty acids, amines, and amino acids.

The lipid and protein composition of a block of cheese remains solid at refrigerator temperatures, similar to unsalted butter. Flavor and odor components are less easily released when the temperature is low. It is generally advised to allow cheeses to reach room temperature before consuming them in order to enhance their flavor and texture. If the cheese is heated beyond 26–32°C (79–90°F), the fats will undergo a process called "sweating out" and transform from a soft state to a completely liquid form (Kapoor & Metzger 2008).

After finishing a meal, a cheeseboard, usually referred to as a cheese course, can be served either prior to or following dessert, or as a substitute for the final course. In Britain, it is common to consume cheese and sweet wines, such as Port, after having dessert. In France, cheese is traditionally paired with a robust red wine before to the serving of dessert. A cheeseboard often comprises a selection of cheeses served alongside cookies, crackers, grapes, almonds, celery, or chutney.

4.3. YOGURT

Yogurt is produced by the process of bacterial fermentation of milk. These bacteria ferment the carbohydrates in milk, resulting in the production of lactic acid. Lactic acid then reacts with the milk protein, contributing to the unique acidic flavor and texture of yogurt. Yogurt is typically derived from bovine milk. Yogurt can be produced using the milk of many animals such as water buffalo, goats, ewes, mares, camels, and yaks (Lee & Lucey, 2010). Homogenizing the milk used is not required. It can be either raw or pasteurized. Each type of milk produces noticeably different results. The microbes *Lactobacillus delbrueckii* subsp, bulgaricus and Streptococcus thermophilus are cultured to produce yogurt. Occasionally, supplementary lactobacilli and bifidobacteria are introduced into yogurt either during or after the fermentation process. Some countries require that yogurt must include at least one million colony-forming units (CFU) of *Lactobacillus* bacteria per milliliter. This rule is applicable in China, as stated by Lee and Lucey in 2010.

Once the bacterial culture is added, the mixture is subjected to a temperature range of 30–45°C (86–113°F) for a duration of four to twelve hours (Lee and Lucey, 2010). Elevated temperatures accelerate the fermentation process, although they also heighten the likelihood of a clumpy consistency or the separation of whey. Whole milk plain yogurt consists of 81% water, 9% protein, 5% fat, and 4% carbs, with 4% of the carbohydrates being sugars. 100 grams contains 406 kilojoules (97 kcal) of nutritious energy. Yogurt contains a moderate quantity of protein, phosphorus, and selenium, providing 14–19% of the daily requirement. However, it is a rich source of vitamin B12 (31% DV) and riboflavin (23% DV) (Lee & Lucey, 2010).

Yogurt can also be sweetened and flavored, or it may come in containers with fruit or fruit jam at the bottom to balance out its inherent tartness. Set-style yogurt and Swiss-style yogurt are the two predominant types of yogurt commonly found in grocery stores. Swiss-style yogurt undergoes swirling before it is packaged, while set-style yogurt is placed in separate containers to solidify. Fruit can be added to enhance its sweetness.

Commercial yogurt often contains a significant amount of sugar or other sweeteners, especially in low-calorie yogurts. In order to enhance the thickness and creaminess of certain yogurts, they may include modified starch, pectin (which is naturally present in fruit), or gelatin. Despite lacking any connection to genuine Swiss yogurt, this particular type of yogurt may be marketed and sold as Swiss-style. Some yogurts, known as "cream line" yogurts, are produced with unhomogenized whole milk, resulting in the cream floating to the surface. Sweetened or flavored yogurt is readily accessible in numerous countries, typically in individual plastic cups (Lee & Lucey, 2010). Typical flavors consist of toffee, honey, vanilla, and other fruits. Yogurt varieties that evoke the taste of desserts, such as cheesecake or chocolate, were increasingly popular in the early 21st century.

Prior to yogurt production, it is necessary to heat milk to a temperature ranging from 30 to 45°C (86 to 113°F). This temperature range ensures that the live microorganisms responsible for fermenting the milk into yogurt remain unharmed (Lee & Lucey, 2010). Following that, particular strains of bacteria, specifically Streptococcus thermophilus and *Lactobacillus* bulgaricus, need to be introduced to the milk. The milk should then be maintained at a warm temperature for a duration of four to twelve hours.

Using milk with a higher solids content than conventional milk will result in firmer yogurt. Powdered milk can be utilized to enhance the concentration of solid components. The yogurt making process creates two significant barriers to pathogen growth: heat and low pH (acidity). Both Lourens-Hattingh and Viljoen (2001) assert that in order to ensure a product's safety, both factors are necessary. The effectiveness of acidity alone has been brought into question by recent outbreaks of food poisoning caused by the acid-tolerant *E. coli*. Pasteurization, a process of applying heat, effectively eliminates *E. coli* by simultaneously eradicating harmful microorganisms and altering the structure of proteins. Due to the ability of the microorganisms responsible for fermenting milk into yogurt to tolerate higher temperatures than most harmful bacteria, the temperature has a dual role in both promoting and hindering the production of yogurt. Optionally, the yogurt can be subjected to straining once it has solidified in order to increase its viscosity and reduce its whey content. Yogurt comes in several forms, such as pasteurized yogurt and probiotic yogurt. Pasteurized yogurt undergoes heat treatment to eradicate germs, whereas probiotic yogurt is created by submitting pasteurized yogurt to a technique that destroys bacteria. Probiotic yogurt contains viable bacterial cultures, such as *Lactobacillus*, which can confer health advantages.

The health benefits of yogurt differ based on the specific variety consumed. Benefits of this include the reduction of symptoms associated with irritable bowel syndrome, the prevention of osteoporosis, and improved digestion. Yogurt is a beneficial dietary source of protein, calcium, vitamins, and probiotics. These components can enhance the composition of gut bacteria and mitigate digestive issues. Low-fat yogurt is an advantageous protein source for individuals adhering to a weight-loss regimen. The presence of probiotics in yogurt has the capacity to augment the body's immune system.

Nevertheless, it is crucial to acknowledge that not all yogurt products provide healthful qualities. Certain yogurt products may contain additional sugar and other additives that may not be suitable for ingestion. It is advisable to select yogurts that are free from added sugar and needless additives. Yogurt in its natural form can be a dietary choice that is both low in calories and high in nutrients, containing a significant amount of protein.

4.4. ICE CREAM

The initial viable technique for producing ice cream was found upon the realization that the addition of salt might decrease the freezing point of fragmented ice. The implementation of the wooden bucket freezer equipped with rotating paddles significantly facilitated the process of creating homemade ice cream. The advent of mechanical refrigeration enabled the widespread distribution of ice cream. Drugstore soda counters and ice cream shops thrived. With the widespread usage of freezers in households, about 50% of all frozen desserts are consumed at home (Goff, 2013).

Ice cream and other frozen desserts are required to comply with strict restrictions. For example, ice cream must have a minimum of 10% fat and 20% total milk solids. During the freezing process, the addition of air can cause the volume to double. However, the condition that the end product must weigh at least 4.5 pounds per gallon prevents the volume from exceeding a 100% increase. The water content is restricted due to the requirement that the total weight of food solids must be 1.6 pounds per gallon (Goff, 2013). According to laws, it is necessary to declare all ingredients, and certain additives, such as stabilizers, can only be utilized in minimal quantities.

The primary categories of frozen desserts include ice cream, frozen custard, ice milk, frozen yogurt, sherbet, and water ices. Ice cream has the highest fat content, with a potential fat content of up to 20%. Ice milk is commonly known as "light" or "reduced-fat" ice cream. The milk contains a fat content ranging from 2 to 7%, and a minimum of 11% total milk solids. Frozen yogurt, a refined frozen dish, shares the same constituents as ice cream. The weight of the substance must be at least five pounds per gallon, and it must have a minimum of 3.25% milk fat and 8.25% milk solids other than fat. The milk fat content in low-fat frozen yogurt typically varies from 0.5 to 2%, as stated by Goff in 2013.

The primary constituents of ice cream include sugar, milk, cream, flavoring, and stabilizer. In order to produce a product at a lower cost, one can utilize more affordable ingredients such as corn syrup, dry whey, and artificial flavorings as substitutes. The initial step in ice cream production is to create a suitable blend. The mixture contains several dairy products such as buttermilk, condensed or dry skim, frozen cream, fresh milk and cream, dairy whey, and whey protein concentrate (Goff, 2013). Sucrose, corn syrup, honey, and other syrups are categorized as sugars. Minimal amounts of stabilizers and emulsifiers help prevent the formation of ice crystals, particularly when there are fluctuations in temperature during storage.

The ice cream mixture is subjected to a minimum temperature of 79°C (175°F) for a duration of 25 seconds. Next, the mixture is heated and then homogenized to guarantee a more even and consistent texture and structure. In addition, the process of homogenizing prevents the mixture from undergoing churning in the freezer, which leads to the separation of fat granules (Goff, 2013). In addition, the process of homogenizing leads to an increase in the viscosity of the mixture. This is due to the fact that smaller fat globules have a larger surface area, which enables the milk protein to absorb more water and result in a more thick and sticky fluid.

After being homogenized, the heated mixture is quickly cooled to a temperature of 4.4°C (40°F) (Goff, 2013). The mixture must undergo a period of aging at this specific temperature for a minimum of four hours to allow the fat to solidify and the fat globules to form clumps. The process of aging enhances the speed of freezing and results in a more seamless final product. There are two methods for freezing the mix: batch freezing, which entails producing one quantity at a time, or continuous freezing, which utilizes a constant flow of mix. Both approaches aim to partially solidify the product while also introducing air. The freezing process is conducted within a cylindrical barrel that is cooled by a refrigerant, which can be either ammonia or Freon.

Stainless steel Dasher blades are inserted into the barrel to remove the frozen mixture from the sides of the freezing cylinder and mix in air to create the final result. The quantity of air introduced during the freezing process is controlled by adjusting the speed of a pump or dasher (Cook & Hartel, 2010). The freezing times for the continuous freezer and batch freezer differ, with the continuous freezer requiring approximately 10 minutes and the batch freezer being instantaneous, depending on the precise conditions.

When partially solidified ice cream is removed from the freezer, its temperature typically ranges from –9 to –5°C (16 to 23°F). Afterward, it is placed in a suitable container and transported to a blast freezer, where the temperatures range from –29 to –34°C (–20 to –30°F) (Cook & Hartel, 2010). The ice cream in this freezer freezes without agitation, as quick freezing at this stage inhibits the development of sizable ice crystals and promotes a velvety consistency and texture. The duration of usage of the hardening chamber depends on the size of the package, often taking 6 to 12 hours to lower the internal temperature of the ice cream to −18°C (0°F).

Figure 4.3. *Ice cream-manufacturer.*

Source: Smith (2022).

Most ice cream manufacturers (Figure 4.3) generate a fundamental blend of milk, cream, sugars, and stabilizers, to which flavors can be incorporated just prior to the freezing process. Various high-volume flavors, such as strawberry, chocolate, and vanilla, can be mixed together in individual large-scale tanks. Flavors with large particles, such as those found in fruit, nuts, cookies, or candy bits, are introduced into the freezer's outlet using a device called a "feeder" (Cook & Hartel, 2010). Selected nuts and fruits undergo meticulous selection and get additional precautions to avoid the introduction of undesirable microbes into the pasteurized mixture.

4.5. CREAM

Cream is a dairy product derived from the fatty layer that is separated from milk prior to homogenization. The fat in unhomogenized milk ascends to the surface over time due to its lower density. Centrifuges, referred to as "separators," are utilized in industrial cream manufacturing to accelerate this procedure (Smiddy et al., 2009). Cream is available in different grades throughout countries, which are determined by the varying levels of butterfat content. It possesses a significant amount of saturated fat and can be dehydrated into a powdered form for transportation to distant markets.

In the dairy business, cream is extracted using mechanical separation. Cream is the portion of milk that contains a high concentration of fat globules that naturally separates and floats to the surface when milk is allowed to settle. Homogenized cream contains smaller fat globules, which makes it less appropriate for the process of whipping (Smiddy et al., 2009).

Various varieties of cream are available, each possessing distinct properties and serving certain purposes. The following items are included:

- *Cream Clot*: This is a British delicacy with a smooth texture, a hint of sweetness, and a minimum milk fat content of 55%. The process involves subjecting cream to prolonged heating in a shallow pan, resulting in the coagulation and ascent of the creamiest portion to the surface.
- *Heavy Cream:* Also referred to as "heavy whipping cream," is a thick mixture with a minimum of 36% milk fat. It is excellent for making homemade whipped cream since it can be whipped into firm peaks.
- *Whipping Cream*: This cream contains a milk fat level that is within the range of 30% to 36%. It is utilized in the creation of desserts and can be manually whipped or dispensed from a pressurized container.
- *Light Cream*: Also referred to as "table cream," contains a milk fat percentage ranging from 18% to 30%. It is effective in both savory and sweet recipes.
- *Sour Cream*: This is a type of cream that has at least 18% milk fat and obtains its somewhat acidic taste through the process of fermentation. The substance has a chilly and almost jelly-like consistency.
- *Half and Half*: This is a well-liked addition to coffee that consists of a blend of equal parts whole milk and light cream, containing a milk fat level that falls between 10.5% and 18%. Its relatively low-fat level prevents it from being whipped.
- *Ice Cream*: This is a frozen dessert that contains at least 10% milk fat and is characterized by its low temperature, sugary taste, smooth consistency, and dense texture.

Commercial sour cream is produced by combining light cream with a butterfat content ranging from 18 to 20%. Following the process of pasteurization, the cream undergoes the introduction of bacteria that actively generate lactic acid. Afterward, the mixture is cooled and allowed to develop for a period of 12 to 48 hours, leading to a tangy taste and a partially solid consistency. Sour half-and-half, with 10.5–18% butterfat cream, is a comparable item (Smiddy et al., 2009). Cream can be extracted from milk or whey, and the term "sweet cream" specifically denotes cream that has been skimmed from milk. Sour cream, a type of cream that has undergone partial fermentation, is widely available in many countries and can be used in a variety of culinary preparations, both sweet and savory. The hue of cream might differ based on the dietary intake of the animals responsible for milk production. Cream derived from cows that graze on natural pasture frequently contains carotenoid pigments, resulting in a faint yellow hue. The cream derived from goat's milk, water buffalo milk, or cows that are fed a diet mostly consisting of grain or grain-based pellets has a white color (Hae Soo et al., 2013).

Cream is a multifunctional component employed in a diverse range of culinary preparations, such as cakes, ice cream, sauces, soups, stews, puddings, and custard bases. Additionally, it is frequently employed as a garnish for ice cream sundaes, milkshakes, lassi, eggnog, and sweet desserts. Indian curries, such as masala dishes, frequently include cream as an ingredient.

Cream can be fortified with stabilizers and thickening agents to improve its texture and stability. Thickeners such as alginic acid, sodium bicarbonate, and tetrasodium pyrophosphate are commonly used.

4.6. CONDENSED MILK

Conventionally, condensed milk is prepared from bovine milk through the process of evaporating a significant amount of water and incorporating sugar. When milk is heated slowly with sugar, it thickens and forms a syrupy substance known as condensed milk. Once it cools, the mixture will undergo thickening and become suitable for usage. If one does not use it immediately, it can be stored in the refrigerator for several weeks.

Around 60% of the water content in cow's milk is removed during the process of producing condensed milk. The words "condensed milk" and "sweetened condensed milk" are often used interchangeably, with the latter typically referring to condensed milk with added sugar (Oliveira, Penna, & Nevarez, 2009). Sweetened condensed milk, when canned and unopened, has a long shelf life and does not require refrigeration for several years. The result is highly viscous and has a pronounced sweetness. The chemical is employed in various dessert recipes in many countries.

4.6.1. Production

Once raw milk has been clarified and standardized to achieve the required fat to solid-not-fat (SNF) ratio, it is subjected to a brief heating process lasting a few seconds, raising its temperature to 85–90°C (185–194°F). During this heating process, certain microorganisms are eradicated, leading to a decrease in fat separation and the prevention of oxidation (Oliveira, Penna, & Nevarez, 2009). Once the water in the milk has evaporated, sugar is gradually added to the evaporated milk in amounts ranging from nine to eleven. The addition of sugar extends the shelf life of sweetened condensed milk. Sucrose increases the osmotic pressure of the liquid, which hinders the growth of microbes. Lactose crystallization is initiated after the sweetened evaporated milk has been cooled.

To produce condensed milk, combine one volume measure of evaporated milk with one and a quarter volume measure of sugar in a pot. Heat and stir the mixture until the sugar is fully dissolved. Then, let it to cool. Alternatively, one can simmer regular milk and sugar together until they reduce by 60% (Oliveira, Penna, & Nevarez, 2009).

Below are instructions for making condensed milk:

- *Step 1:* Pour milk and sugar into a saucepan.
- *Step 2:* Proceed to simmer the milk until it reaches a thicker consistency. During the process, make sure to stir frequently. It is important that the food does not scorch or clump together at the bottom of the pan.
- *Step 3:* Extinguish the flame. Observe the enchanting transformation that occurs when a minute quantity of baking soda is added to your pan.
- *Step 4:* Strain your milk through a strainer if you notice any clumps in it.

4.6.2. The Health Advantages of Condensed Milk

Condensed milk contains a higher amount of sugar compared to regular milk, hence amplifying its advantages. Condensed milk is a rich source of essential minerals such as calcium, vitamin D, vitamin B12, vitamin A, riboflavin, phosphorus, and protein. The presence of these nutrients (Oliveira, Penna, & Nevarez, 2009) is crucial for maintaining strong bones and teeth, a well-functioning immune system, and a properly functioning cardiovascular system.

- **Possibly Reduce Cholesterol Levels:** Milk and dairy products containing cholesterol and saturated fats may increase the likelihood of developing cardiovascular diseases. According to a study, the

Maasais, an African tribe that eats milk, exhibit a low prevalence of cholesterol. Dairy products are believed to include indigestible carbohydrates, sometimes known as sugars, that are metabolized by the intestinal microbiota through fermentation. This intervention decreases blood cholesterol levels and alters the synthesis of cholesterol (Jouki et al., 2021).

- **High in Calories:** The high caloric density of a little quantity of sweetened condensed milk might have either positive or negative effects, depending on the specific needs or requirements. It can be a beneficial instrument for persons seeking to increase their weight, but it may provide excessive calories to those attempting to reduce their weight. Condensed milk is primarily utilized as a component in confectionery rather than being ingested independently or as a dietary supplement, so there is limited information available regarding its health advantages.

4.6.3. Risks of Condensed Milk

Although condensed milk may have similar nutritional advantages as whole milk, it is not a good substitute for it. Due to the presence of added sugar and high calorie content, condensed milk may not be the optimal option for individuals aiming to lose weight. If reducing calorie intake is a priority, then evaporated milk is a preferable option. Individuals with lactose intolerance may have adverse consequences due to the presence of lactose in condensed milk. Possible symptoms include gas, cramps, and abdominal discomfort (Jouki et al., 2021). Condensed milk contains a higher amount of lactose compared to regular milk since it is more concentrated in the same volume. Although sweetened condensed milk is calorie-dense, it also has elevated levels of saturated fat, which can be detrimental to individuals with heart conditions or diabetes. Studies suggest that having a diet high in saturated fats can elevate insulin levels and low-density lipoprotein (LDL) cholesterol. Elevated levels of LDL cholesterol increase the likelihood of developing heart issues. Moreover, it can exacerbate insulin resistance, a condition that hinders the body's ability to metabolize the sugar responsible for the development of type 2 diabetes (Jouki et al., 2021).

Sweetened condensed milk can be stored in the refrigerator or freezer for an extended period of time, typically many months. Upon opening, it should be stored in a refrigerated environment and kept in a container that is airtight. If stored in the refrigerator, it will remain fresh for a duration of two to three weeks. Factory-sealed cans of sweetened condensed milk typically remain fresh for up to one year beyond the indicated expiration date. According to Jouki et al.

(2021), in order to have an extended shelf life, it is crucial to have a high sugar content and hermetically sealed cans. The shelf life significantly decreases once it is opened, so make sure to use it within a few days.

4.7. WHEY PRODUCTS

Whey is the residual liquid that remains after milk has undergone the process of straining and curdling. It is a derivative of casein or cheese manufacture and has various industrial applications. Sweet whey is a secondary product that is obtained when producing rennet-based hard cheeses like Swiss or cheddar (Anand et al., 2013). Sour whey, also known as acid whey, is a residual substance derived from the manufacturing process of acid-based dairy products, such as strained yogurt. Whey proteins consist of bovine serum albumin, heavy and light chain immunoglobulins, β-lactoglobulin (48%–58%), α-lactalbumin (13%–19%), and other less significant whey proteins (Anand et al., 2013).

Essentially, dairy products such as milk and whey consist of three main components: lactose, soluble and insoluble proteins, and lipids, which are suspended in the product.

Whey, a byproduct of the dairy industry, contains significant amounts of essential components, specifically soluble proteins such as immunoglobulin, β-lactoglobulin, α-lactalbumin, lactoferrin, and lactoperoxidase. They are frequently utilized as food ingredients, typically in a dehydrated state, in various food formulations including bakery, confectionery, health, and sports supplements. Whey products possess significant biological properties, including antibacterial, anticarcinogenic, and immunomodulatory activity. They also exhibit functional characteristics such as gelation, foaming, and emulsifying properties. Additionally, whey products are nutritionally valuable due to their high content of essential amino acids, which contribute to overall health (Anand et al., 2013).

Whey is the liquid that remains after the proteins in milk have been separated. Microorganisms, such as those found in whey from cheese, the inclusion of acid for acid casein production, or the employment of enzymes for rennet casein manufacturing, can all facilitate the precipitation process. Whey is classified into two distinct categories: acid whey and sweet whey. Acid whey is a byproduct that is generated during the manufacturing process of acid casein and cottage cheese (Anand et al., 2013). The term "sweet whey" encompasses all other types of whey. The majority of whey produced worldwide consists of sweet whey. In the United States, approximately 94% of the total whey produced consists of sweet whey, while the remaining 6% is comprised of acid whey. Whey is a weakened source of nutrients. The composition of the contents is around 0.8%

protein, 4.5% lactose, 0.7% minerals, and 94% water (with 6% total solids). Drying this diluted stream can be done to produce whey powder or sweet whey powder (Anand et al., 2013). Reduced lactose whey powder can be obtained by decreasing the lactose content in the whey and subsequently dehydrating the resulting product. Similarly, techniques such as electrodialysis can decrease the mineral content of whey, resulting in the production of demineralized whey. This demineralized whey is then dried to create demineralized whey powder. The next phase in the whey processing process involves concentrating the proteins by removing lactose and minerals.

Whey protein concentration refers to powders containing a minimum protein content of 25% or more. Whey protein concentrates generally have protein contents of 35%, 55%, and 80%. According to Anand et al. (2013), whey protein isolates that contain above 90% protein when measured without water are created by an extremely thorough purification process. The protein concentration yields lactose and calcium-rich milk minerals as byproducts. The markets for nutritional supplements and pharmaceuticals are increasingly utilizing these by-products for various purposes. Lactose is mostly used as an excipient in the tableting process, whereas milk minerals are employed to enhance the calcium content in various food and beverage products.

Whey, among other consumable products, is utilized in the production of whey cheeses such as ricotta, Norwegian brunost, and whey butter. Whey has a low-fat content. To produce a few pounds of whey butter, it requires 1,000 pounds of whey. Moreover, it is used as a component in various processed products, including commercial pastries, breads, and crackers, as well as animal feed (Kilara, 2015). β-lactoglobulin and α-lactalbumin are the primary constituents of whey proteins. Sweet whey contains glycomacropeptide (GMP). Additionally, it contains a substantial amount of lactose, which can be harnessed to produce beneficial chemicals derived from lactose. There are multiple use for dairy whey that remains after producing cheese at home. It serves as a dough conditioner and can substitute skim milk in most recipes for baked items that require milk.

The primary constituent of most protein powders, commonly used by bodybuilders and athletes to meet their daily protein needs for muscle development and upkeep, is whey (Kilara, 2015). Leucine, a branched-chain amino acid, is highly present in whey protein, making it ideal for promoting muscle growth and facilitating tissue repair. Whey, like all types of milk, undergoes pasteurization to prevent the growth of harmful bacteria. The temperature is raised to a range of 70 to 80°C (158 to 176°F), and thereafter lowered to 4°C (39°F). Studies have shown that subjecting bacteria to extremely high temperatures during this method results in a 99.7% mortality rate, without allowing the protein to solidify into a mass (Kilara, 2015).

CHAPTER 5

THE CHEMISTRY OF FLAVOR IN DAIRY PRODUCTS

Contents

Milk and milk derivatives are a substantial element of everyday diets in numerous regions around the globe. The flavor of milk and milk products is the main determinant of consumer acceptability and marketing, together with their nutritional content. An expert can rapidly ascertain the freshness of milk by evaluating its taste. While previous study has explored the flavor chemistry of milk and milk products, there is still a significant amount of knowledge to be gained regarding the flavor chemistry and biochemistry of dairy products. Researchers specializing in flavor and sensory perception continue to face challenges in accurately characterizing the flavor of dairy products (McGorrin, 2001).

Figure 5.1. *Dairy processors offer diverse flavors to satisfy consumers.*

Source: Canning (2020).

The flavor of dairy products is a crucial feature that affects their shelf life, consumer acceptance, and other aspects. The taste of liquid milk is mostly influenced by the quality of the unpasteurized milk, which includes the quantity of heat-resistant enzymes and microbial counts. Heat treatment is employed to decrease or remove the presence of microorganisms and enzymes in milk, resulting in a distinct flavor that differs from raw milk (Figure 5.1). This process ensures safety and extends the milk's shelf life. In the United States, milk is commonly subjected to HTST pasteurization, which involves heating it to a minimum temperature of 72°C for 15 seconds. This process extends the shelf life of milk to around 3 weeks (McGorrin, 2001). Thermal treatments, including UHT and ultra-pasteurization (UP), are employed to extend the longevity of milk.

The Maillard reactions between heat-treated lactose and amino groups in milk proteins result in the formation of several flavor compounds, such as maltol and diacetyl, as well as molecules containing sulfur and nitrogen. Methyl ketones are formed as a result of the β-oxidation of free fatty acids, which occurs during the degradation of lipids in milk during thermal processing (McGorrin, 2001). The process of heat denaturation leads to the release of volatile sulfur compounds from β-lactoglobulin in serum protein. This occurs as the protein unfolds and exposes reactive sulfides.

5.1. THE MAILLARD REACTION IN DAIRY PRODUCTS

To ensure microbiological safety before being sold and consumed, milk is commonly subjected to heat treatment. There are three distinct types of heat treatment: ultra-high temperature (UHT) treatment, high temperature short time (HTST) pasteurization, and low temperature long time (LTLT) pasteurization (Arena et al., 2017). The Maillard reaction occurs in milk during all heat treatments (Figure 5.2).

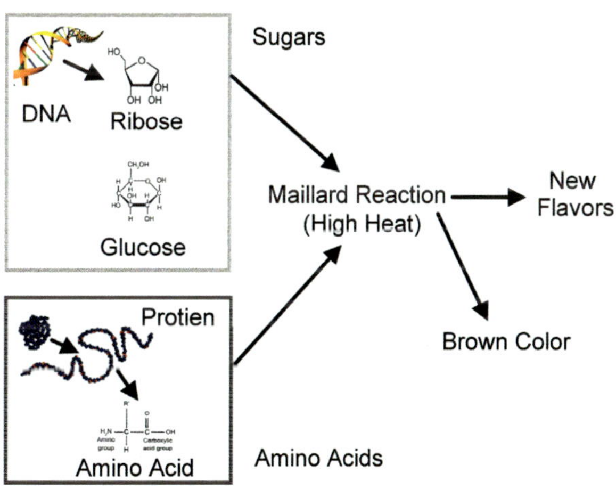

Figure 5.2. *The Maillard reaction and caramelization.*

Source: MySweetDelights (2013).

The Maillard reaction resulted in a loss of lysine ranging from 2.5% (in cooked cream) to 36.2% (in condensed milk). The reaction described has a notable effect on the amount of lysine needed on a daily basis. This influence varies from 13% (in milk pills and soft nougat) to 61% (in dietetic meals). The Maillard reaction, which does not include enzymatic glycation, is a very intricate chemical reaction between an amino group and a carbonyl group. This reaction commonly takes place during the processing or storage of food. When it comes to milk, lactose undergoes the early, intermediate, and advanced stages of the Maillard reaction. This reaction occurs when lactose reacts with the free amino acid side chains of milk proteins, namely the ε-amino group of the lysine residue. As a result, a diverse range of Maillard reaction products are formed. Multiple research have investigated the interactions between lactose and milk proteins, revealing evidence of the formation of various Maillard reaction products in heated milk.

The Maillard reaction progresses from the formation of an Amadori product to either the 3-deoxyosone or 1-deoxyosone pathway, depending on the pH of the process. A third reaction pathway exists in the Maillard reaction of disaccharides such as lactose, as demonstrated by Arena, Renzone et al. in 2017. The 4-deoxyosone route is the subject of discussion. Lactose is the main carbohydrate present in milk. Thus, the aforementioned three channels are the means by which the Maillard process in milk progresses. Ultimately, the Maillard reaction leads to the formation of melanoidins, which are compounds responsible for browning.

The Maillard reaction's influence on milk proteins: The Maillard process has several effects on milk proteins, such as altering their bioavailability, solubility, ability to form structures, capacity to emulsify, and stability when exposed to heat. Moreover, the Maillard reaction occurring between lactose and milk proteins leads to the production of flavor and color compounds.

During the Maillard process, lactose commonly has a reaction with the ε-amino group of the lysine residue found in milk proteins. Consequently, the loss of lysine in the Maillard reaction increases with the intensity of heat treatment. The modified lysine is no longer available as a source of nourishment. As an example, the steam injection method of direct heating caused 3.6% and 6.8% of the lysine in whole milk to become blocked at temperatures of 120°C for 400 and 290 seconds, respectively (Arena et al., 2017). The lysine content increased from 11.0% to 13.0% after subjecting it to indirect heating at 115°C for a duration of 10 to 40 minutes.

Additionally, the glycated β-lactoglobulin shown increased stability when exposed to heat and acidic pH. Moreover, the process of glycation improved the ability of β-lactoglobulin to form and emulsify substances. The findings indicated that the Maillard process plays a significant role in imparting unique features to milk proteins.

The Maillard reaction exerts a substantial influence on the functionality of milk proteins and the sensory attributes of milk and dairy products. Specifically, an excessive advancement of the Maillard reaction and the synthesis of melanoidins during the milk production process are undesirable since they greatly diminish the economic value of the milk. Therefore, it is essential to identify the byproducts of the Maillard process in order to preserve the quality of milk. Several heat-induced markers have been suggested thus far to oversee and control the heat treatment administered to dairy products and milk. Specifically, research has found that the most favorable indicators are the levels of furosine, hydroxymethylfurfural (HMF), and lactulose. These concentrations increase following the application of heat treatment (Newton et al., 2012).

The Maillard reaction is responsible for the various colors and flavors observed in food, such as the browning of grilled or seared meats, the umami flavor and browning of fried onions, and the roasting of coffee. It enhances the hue and taste of dehydrated and concentrated milk, caramel, aged garlic, cocoa, toasted marshmallows, and roasted peanuts (Newton et al., 2012). Additionally, it causes the crust of baked goods to become darker and imparts the typical golden-brown color to French fries and other crispy foods.

The Maillard process is initiated by the condensation interaction between an amino group and the carbonyl group of a reducing sugar, leading to a sequence of interconnected processes. In dairy systems, the lactose hydrolyzed products contain glucose and galactose, which act as the reducing sugar. The predominant amino groups in question are primarily derived from lysine and arginine, as well as the N-terminal amino groups found in proteins and peptides. In order to organize this reaction in a systematic manner, the Maillard reaction is typically divided into initial, intermediate, and final stages. The process of combining an amino group with a carbonyl group to form a Schiff's base, which then undergoes a rearrangement to produce the Heyns or Amadori product, is referred to as the early Maillard reaction (Newton et al., 2012).

The Amadori product undergoes degradation during the intermediate stage of the Maillard reaction, resulting in the release of the amino group and sugar fragmentation products, including highly reactive α-dicarbonyl molecules (Newton et al., 2012). The Maillard reaction is a complex process that involves condensation, dehydration, fragmentation, and polymerization reactions. These reactions lead to the formation of advanced glycation end (AGE) products and, ultimately, melanoidins. Melanoidins are large polymerized molecules that are responsible for color changes and browning.

5.2. LIPID OXIDATION IN DAIRY PRODUCTS

Oxygen molecules react with unsaturated fat molecules during the process of lipid oxidation. Undesirable in food items, particularly milk fats, lipid oxidation can lead to the formation of volatile taste compounds, potentially hazardous oxidation products, and a general deterioration in the quality of the fat (García-

Martínez & Marquez-Ruiz, 2009). Unsaturated fat oxidation typically follows a three-stage autocatalytic free-radical mechanism, which includes initiation, propagation, and termination. It leads to the formation of unstable intermediate fatty acid hydroperoxides, which eventually decompose and generate various volatile compounds that add to the rancidity of the fat. The rate and extent of lipid oxidation in milk and milk products are influenced by various factors, including oxygen, light, both endogenous and exogenous metals, antioxidants such as ascorbic acid, tocopherols, carotenoids, thiols, proteins, enzymes, components of the milk fat globule membrane, and storage temperature (García-Martínez & Marquez-Ruiz, 2009). Cholesterol oxidation in milk fat can lead to the creation of cholesterol oxidation products (COPs). Although milk and milk products often do not contain significant levels of COPs, it is advisable to optimize processing conditions to minimize the chances of COP generation.

Food lipid oxidation is a significant concern since it leads to the degradation of the product's quality, causing alterations in both its texture and appearance, as well as its nutritional composition. Unsaturated fatty acids, commonly present in phospholipids and triacylglycerols, ultimately degrade into volatile compounds in food, resulting in an unpleasant odor and rancidity. Fatty acids are more susceptible to lipid oxidation due to the decrease in bond dissociation energy of carbons with methylene interruptions as unsaturation increases (García-Martínez & Marquez-Ruiz, 2009). Adhering to the latest Dietary Guidelines, which promote the adoption of polyunsaturated fatty acids as a superior source of fats, is becoming more difficult.

While it is understandable that accelerated shelf-life tests are conducted to save time, it is important to note that temperature greatly affects oxidation reactions. The effects observed at high temperatures may not necessarily be applicable to shelf life at room temperature due to variations in oxygen solubility, partial pressure, and the formation of side products such as caramelization and Maillard reactions. Reactive oxygen species play a role in the oxidation of proteins and lipids, which are likely to be present in protein-stabilized emulsions. Metal ions can expedite these reactions and lead to the creation of various oxidation byproducts, such as alcohols, ketones, and aldehydes, along with hydroperoxides, which are the primary outcome of oxidation.

To evaluate the effect of milk fat components on glycation at a temperature of 120°C for a duration of 60 minutes, we analyzed the levels of glycation products and lipid oxidation products in three groups: skim milk, whole milk, and a simulated milk fat group. The number of glycation sites in α- and β-casein, the levels of advanced glycation end products (AGEs), and the rate of growth of carbonyl compounds are all higher in whole milk compared to skim milk. This indicates that milk fat significantly enhances protein glycation (García-Martínez & Marquez-Ruiz, 2009). The formation of glycation products was promoted in the milk fat simulation groups by the proportional incorporation of oleic acid and linoleic acid (LA) into milk fat in skim milk; palmitic acid did not have a similar effect.

Los Angeles displayed strong support for the advancement of AGEs. The key factors involved in milk glycation include the production of carbonyl compounds, the condensation of protein carbonyl amines, and the generation of lipid oxidation radicals. Thus, it is postulated that free radicals produced during fat oxidation are responsible for the early glycation process. Additionally, the carbonyl compounds resulting from fat oxidation are believed to have important roles in milk glycation by serving as key intermediates in the glycation process. These roles are influenced by factors such as the type of fat, its structure, and the rate at which it breaks down (Kitts & Tomiuk, 2013).

The primary factor responsible for oxidative changes in dairy products is the presence of metallic impurities, particularly copper contamination. With the enhancement of the products' hygienic quality, it is more imperative to exercise control over copper contamination during the processing stage. However, it is not necessary for all milk to be tainted with copper in order to have a flavor that is affected by oxidation (Kitts & Tomiuk. 2013). Only a minor fraction of milk naturally develops an oxidized taste; when this milk is mixed with regular milk, its flavor is unaffected. Consequently, its commercial importance is lower compared to milk, which is more susceptible to the development of a metallic, oxidized taste.

Studies have shown a connection between reduced levels of microorganisms in milk and the development of an oxidized taste. Milk loses its sensitivity when exposed to copper because bacteria multiply and either deplete the oxygen in the milk or lower its potential. The presence of an enzyme in milk was first believed to be responsible for the development of the oxidized flavor, as it promoted the oxidation of fat, a process helped by copper. A recent study on the flavor of cooked food has cast doubt on the role of an enzyme in generating the lack of flavor. This study supports the prevailing idea in literature that the phospholipid fraction is responsible for the flavor (Kitts & Tomiuk, 2013).

Research has demonstrated that oxygen and light have substantial impacts on facilitating oxidative modifications. Oxidation can still occur even without light, albeit at a relatively low pace. The oxygen can originate from either a blended or unbound source. Both severe acid and alkaline reactions have been shown to promote oxidative changes. Butter made from high-acid cream that has been polluted with copper and salt can have a fishy odor when refrigerated. Over-neutralization has been found to stimulate oxidative modifications, as reported by Kitts and Tomiuk in 2013. Compared to metallic pollution, the influence of moisture and salt on the occurrence of oxidative changes appears to be insignificant.

5.3. ENZYMATIC REACTIONS IN DAIRY PRODUCTS

Enzymes have the potential to induce undesired alterations in milk and dairy products. For instance, lipase enzymes can induce hydrolytic rancidity in milk and dairy products. Plasmin can induce bitterness and/or trigger age gelation in milk that has undergone ultra-high temperature (UHT) processing (Figure 5.3). Lecithinase can induce the formation of small particles in cream, resulting in a bitty texture.

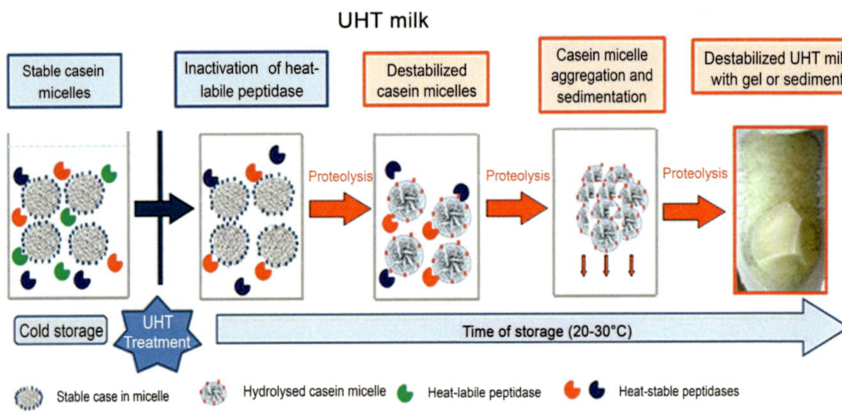

Figure 5.3. *UHT milk destabilization due to proteolysis during storage.*

Source: Machado et al. (2017).

Milk contains a wide range of naturally occurring enzymes, including antimicrobial enzymes, which play different roles, have different levels of stability during processing, and have variable impacts on dairy products. These enzymes are also important for ensuring consumer safety. Certain enzymes, such as lactoperoxidase, possess advantageous properties, while others, like alkaline phosphatase, serve as indicators for processing. Additionally, enzymes like plasmin and lipoprotein lipase can either positively or negatively affect the quality of dairy products, depending on the specific product in question (Silanikove & Shapiro, 2012). The investigation of milk's enzymes is a vital field of knowledge that is currently under active research in the disciplines of dairy science and biochemistry. Milk can be transformed into various products using enzymes, including cheese, kefir, cream, yogurt, butter, and lactose-free products. Enzymes also enhance milk production in animals and contribute to the preservation and shelf life of milk. They improve taste, consistency, and offer health benefits. Enzymes are used to reduce the BOD (Biochemical Oxygen Demand) and COD (Chemical Oxygen Demand) of dairy effluents, which helps protect the environment and allows for the extraction of valuable goods from dairy waste.

The primary objectives of research should be to uncover novel enzymes, optimize their synthesis, and investigate their properties to enhance our understanding of their function and potential in the dairy product manufacturing process (Silanikove & Shapiro, 2012). Recombinant DNA technology can be applied to enhance dairy enzymology by increasing the catalytic activity and stability of enzymes by overexpression or modification. The utilization of enzyme technology is expected to increase due to the nutritional benefits of dairy products, growing consumer awareness, and a trend towards low-fat, low-lactose, or healthy diets. There will be a continual increase in the number of enzymes in the future, and all aspects of the dairy industry will experience a substantial increase in the utilization of these enzymes.

5.3.1. The Significance of Enzymes in Milk

Milk has a diverse range of biological components, such as growth hormones, enzymes, functional peptides, oligosaccharides, and immunoglobulins, in addition to being rich in proteins, lipids, carbohydrates, minerals, and vitamins. Milk from mammals contains around sixty-three naturally occurring enzymes. Enzymes like lysozyme, catalase, superoxide dismutase, lactoperoxidase, xanthine oxidase, ribonuclease, and others play a crucial role in maintaining the stability of milk and protecting individuals from harmful pathogens. This is due to their antioxidant and antibacterial properties (Silanikove & Shapiro, 2012).

Milk is highly susceptible to degradation due to microbial activity and the presence of various enzymes. Enzymes can cause biochemical changes in both raw and processed milk, leading to reduced stability and shorter storage duration. Exogenous enzymes are used to modify milk, while indigenous enzymes have technological uses. Certain enzymes, such as sulfhydryl oxidase and superoxide dismutase, play a crucial role in preserving milk quality. However, enzymes like lipase, proteinase, acid phosphatase, and xanthine oxidoreductase can negatively impact milk quality. Enzymes like lactoperoxidase and lysozyme have antibacterial activity, while specific enzymes contribute to the flavor of cheese during ripening. However, some enzymes, like lipase, can cause undesired alterations in milk and dairy products, such as hydrolytic rancidity. Plasmin can induce bitterness and/or trigger age gelation in ultra-high temperature (UHT) treated milk. Lecithinase can induce the formation of small particles in cream, resulting in a gritty texture.

During the process of metabolism, the β-glycosidic link in D-lactose is hydrolyzed, resulting in the production of D-galactose and D-glucose. These chemicals have the ability to traverse the gut walls and get access to the circulation. Lactase catalyzes the following general reaction: The reaction of lactose with water produces β-D-galactose and D-glucose. Similarly, the reaction of C12H22O11 with water produces C6H12O6 and C6H12O6, along with the release of heat (Fox, 2003).

The catalytic process of D-lactose hydrolysis maintains the anomeric structure of the substrate products. The precise mechanism by which stereochemical retention is achieved remains unclear, however it is known to require a two-step displacement process. Studies on E. Col lactase have indicated that the enzyme's glutamate nucleophile initiates the hydrolysis process by attacking the galactosyl carbon in the β-glycosidic bond from the axial side (Fox, 2003). Mg-dependent acid catalysis is a potential mechanism for the removal of the D-glucose leaving group. Water hydrolyzes the equatorial nucleophile domain of the enzyme, causing the release of the α-galactosyl group and the formation of D-galactose.

5.4. THE PRESENCE OF AROMATIC COMPOUNDS IN DAIRY PRODUCTS

An aroma compound is a chemical substance that possesses a distinct smell. It is commonly known as an odorant, aroma, fragrance, or flavoring. In order for a chemical or group of chemical compounds to produce a scent, it must have enough volatility to be able to move through the air and reach the olfactory system in the upper part of the nose. For instance, various aromatic fruits possess distinct fragrance components. Strawberries, cultivated for their fragrance in a commercial environment, possess several scent components (Tunick, 2014).

Aromatic compounds, sometimes referred to as arenes or aromatics, are best exemplified by conjugated planar ring systems that possess delocalized pi-electron clouds instead of distinct alternating double and single bonds.

The scent of cheese is mostly attributed to its volatile components, such as fatty acids, alcohols, aldehydes, esters, ketones, lactones, sulfur compounds, amines, and other flavor-active molecules.

In addition to lactic acid, the primary components responsible for the distinctive aroma of yogurt are acetaldehyde, diacetyl, acetoin, acetone, and 2-butanone. The prolonged storage of yogurt results in an undesirable taste, primarily due to the oxidation of lipids. This process leads to the formation of undesirable fatty acids and aldehydes.

The acceptability of yogurt goods by consumers is greatly influenced by the smooth, viscous gel-like texture and sharp acidity that is coupled with a green apple scent. The conventional flavor of yogurt is a result of the harmonious blend of taste and scent that often arises during the process of lactic acid fermentation facilitated by the yogurt starter cultures (Tunick, 2014). The yogurt starting cultures convert lactose and other nutrients in milk into several molecules during fermentation, leading to the creation of diverse flavor and odor compounds. Over 100 distinct volatile chemicals have been documented and generated by various yogurt cultures. A variety of advanced methods and equipment have been employed to detect the volatile chemicals present in yogurt products.

The initial microorganisms are crucial for the development of the cheese's flavor. The main origin of taste in cultured milks and fresh cheeses is the transformation of lactose into lactic acid by lactic acid bacteria, along with the creation of diacetyl and acetaldehyde. The initial bacteria in mature cheeses rapidly perish, and the speed at which they disintegrate and release their enzymes into the system impacts the rate of production of free amino acids. Rennet acts independently to synthesize large, medium, and small peptides. Without the presence of additional enzyme interactions, it can only generate measurable quantities of methionine, histidine, glycine, serine, and glutamic acid (Tunick, 2014). Microbial peptidase activity is the main reason for the presence of free amino acids in cheddar cheese.

These amino acids act as substrates for secondary microorganisms, which frequently determine the type of cheese, along with the results of glycolysis. Additionally, they function as substrates for enzymes obtained from milk. For instance, it seems that milk enzymes are essential for the production of H2S. Rennet liberates methionine, which is subsequently metabolized by starter enzymes to generate methanethiol, a crucial constituent of cheese flavor that potentially enhances the intensity of other flavors. α-Dicarbonyls, such as methylglyoxal and diacetyl, together with the bacteria capable of producing them, appear to play a crucial role in the formation of cheese flavor. This includes the desired flavor of full-fat cheese and the undesirable meaty and brothy flavor seen in low-fat cheese (Tunick, 2014).

The importance of flavor on consumer food acceptability is widely acknowledged. If the food is rejected because of its flavor, it does not have any nutritional value. Regular milk has a mild flavor, but it is not unpleasant. Nevertheless, there are some who have a negative opinion of its understated and unique taste, its texture in the mouth, or what they perceive as an aftertaste. While unfortunate, this situation is comparable to the varying reactions people have towards various types of food (Gierczynski et al., 2011). Given milk's copious nutritional composition and its ability to aid in weight management, it is desirable that individuals who hold a negative stance towards it would be open to reconsidering their position. Some foods require an acquired taste.

The diversity in milk flavor is another notable characteristic. This is primarily due to its lack of flavor, which makes any newly added flavor stand out prominently. In addition, milk can undergo an undesirable change in flavor if it is subjected to heat, light, or air. Aberrations associated with the bovine's nourishment or metabolic processes may also lead to milk with an unusual flavor. Experiments have shown that cows can absorb the flavor of onions into their milk in less than a minute. The dairy industry's exceptional ability to manage the intricate challenge of milk's flavor is truly extraordinary (Gierczynski et al., 2011).

Ethyl hexanoate and ethyl butanoate were the two most influential chemicals in terms of odor in raw bovine milk. It is important to observe that raw milk consistently contains much higher quantities of ethyl butanoate and ethyl hexanoate compared to pasteurized milk (Gierczynski et al., 2011).

5.5. SENSORY EVALUATION OF DAIRY PRODUCTS

The main focus of sensory analysis for dried dairy products lies on evaluating the product's visual appearance and fragrance. Examiners seek a pristine milk aroma devoid of any notable imperfections. These foods may exhibit intermittent aromas of caramelization or grassiness, which could indicate potential problems during the processing stage.

The sensory quality of food items has been considered a vital element since the inception of the food industrialization process due to its impact on the overall quality of the product. The olfactory, gustatory, retro-nasal, tactile, and visual attributes of a food are the primary factors that customers consider when deciding whether or not to approve of it. Quality can be defined as the preference of the consumer, with the extent of positive attributes and absence of negative qualities serving as better indicators of quality ratings. Hence, while selecting and consuming a food product, the primary consideration should be its palatability, or eating quality. Other criteria such as nutrition and wholesomeness are of secondary importance (Clark et al., 2009).

Sensory analysis is used to quantify and characterize the sensory properties of products. Sensory analysis refers to the process of describing and quantifying a product's attributes as perceived by the senses, including sight, sound, smell, taste, and touch. By understanding sensory data, one can give precise directions for food product creation, such as emphasizing a specific attribute during evaluations. This decision-making process involves taking into account economic and ingredient processing factors (Clark et al., 2009). The description of food can extend beyond mere "tasting" and include elements such as color, texture, flavor, aftertaste, aroma, tactile response, and even auditory perception. The phrases "sensory evaluation" and "analysis" are sometimes used interchangeably.

The purpose of sensory assessment is to provide reliable and precise data to the research, production, and marketing departments, enabling management to make informed business decisions regarding the perceived sensory characteristics of the product. By establishing a connection between sensory qualities and instrumental, physical, or chemical analysis, it is possible to achieve cost savings.

Sensory evaluation can be applied in the dairy industry across various domains.

- Analysis of the unprocessed components.
- The development of novel products or the improvement of existing ones through the alteration of ingredients.
- Evaluating the product in relation to its competitors.
- Cost reduction.
- Engaging in the process of monitoring and evaluating the level of quality assurance.
- Identify discrepancies among products from multiple batches or production cycles.
- Choosing appropriate materials for packaging.
- Present newly acquired products to the marketing team.
- Promote the acquisition of new or redesigned products.
- Research on the duration of product viability.
- Establishing links between analytical, instrumental, and sensory aspects.

The sensory evaluation method for dairy products includes a thorough investigation and interpretation of the product's primary sensory aspects (Clark et al., 2009). There are three methods for identifying the reasons of sensory impairments in dairy products: (1) employing chemical processes; (2) conducting microbiological testing; and (3) utilizing sensory evaluation. The sensory evaluation method is the simplest, most efficient, and most direct. Prior to implementing remedial measures during the stages of manufacturing, processing, and distribution, it is essential to accurately identify the nature and causes of sensory impairments. Usually, a small number of knowledgeable "experts" evaluate and assess dairy products by assigning them quality scores. The characteristics of aesthetic, flavor, and texture are evaluated based on the presence or absence of predetermined defects (Cooper, 2017).

Food products can possess either external or intrinsic sensory characteristics. Intrinsic qualities refer to the tangible attributes that customers can observe and often use as a measure of quality, without the necessity for real consumption or usage. The qualities of milk products, such as appearance, color, shape, size, and structure, are all essential factors. The physical attributes of a product are consistently linked to its inherent qualities (Cooper, 2017). Product features that are utilized to assess a product but are not physically present in it are known as extrinsic quality cues. Some examples of these characteristics encompass the cost, brand, production and nutritional details, packaging aesthetics, country of production, seller, and convenience.

The main sensory characteristics linked to dairy products encompass flavor, consistency, texture, hue, and visual presentation. The human senses are employed to discern specific attributes of food. The five fundamental senses involved in sensory perception are gustation (mouth), tactition (skin), olfaction (nose), vision (eyes), and audition (ear). The human sensory system has five primary senses: vision, audition, tactile perception, gustation, and olfaction, which are employed to perceive external stimuli. Taste and smell are considered the most fundamental of all human senses. Chemistry and pain senses, along with smell, respond to chemical stimuli. Human sensory perception encompasses not just the ability to detect heat and cold, but also the sensations of pain, visceral hunger, thirst, fatigue, and balance (Cooper, 2017).

The visual attractiveness of a dairy product can be evaluated by examining its appearance, tidiness, and hygiene, as well as the effectiveness of the bottle or pouch closures in ensuring safety. It is assigned a rather low number ranking on the scoresheet. The presence of imperfections in visual products is associated with individual preferences, with sight being used as a means of evaluation.

5.6. FACTORS AFFECTING FLAVOR OF DAIRY PRODUCTS

Various factors can impact the flavor of dairy products, especially milk. The primary determinant is the excellence of the initial milk, encompassing its absence of impurities, unwanted enzymes, putrefaction microorganisms, and unpleasant tastes. Flavor problems might arise as a result of an unfavorable storage condition.

The taste of milk can also be influenced by processing factors, such as the specific kind of pasteurization employed. Furthermore, factors related to animal management, such as breed, lactation stage, and diet, might influence the taste of milk.

The acceptability of food by customers is greatly impacted by its flavor. If a product is rejected because of its flavor, it does not give any nutritional benefit. Although regular milk is often described as having a mild and unique taste, there are those who do not enjoy its faint flavor, the texture it creates in the tongue, or what they perceive as an aftertaste. Nevertheless, due to milk's copious nutritional composition and its capacity to assist individuals in managing their weight, it is anticipated that those who hold a negative stance towards it will remain open to reconsidering their position. Milk is sometimes regarded as having a lack of distinct taste, which poses a challenge for consumers in defining its flavor. Dairy scientists have dedicated a greater amount of time and effort to identifying the specific molecules responsible for undesirable tastes in milk, rather than focusing on establishing the chemicals that are most essential for enhancing its flavor. High-quality processed milk exhibits a mild taste; however, it also possesses a discernible flavor that has been described as having a subtle combination of saltiness and sweetness. Lactose and milk salts are probable contributors to a portion of this flavor, but there are more factors that contribute to the distinctive taste of milk that may not be immediately apparent.

The sensory experience of milk is significantly impacted by the mouthfeel, which is determined by the emulsification of fat globules in a protein phase that is dispersed in water. The volatile compounds in milk mostly contribute to its fragrance and flavor. Dimethyl sulfide was among the initial notable chemicals discovered in milk that contributed to its aroma. Subsequent to that time, there have been notable progressions in analytical techniques and apparatus, resulting in the detection of numerous additional foul-smelling substances in milk, including aldehydes, ketones, alcohols, fatty acids, lactones, esters, sulfur compounds, nitrogen compounds, and aromatic hydrocarbons.

The complexity in determining the primary odorants responsible for the flavor of milk is due to the abundance of volatile chemicals present in milk, their low levels of concentration, and their low threshold values for being detected by the sense of smell. Scientists have employed sophisticated methods to detect neutral volatile compounds in unprocessed milks from diverse animals. Their findings suggest that the distinct flavors and aromas of milk are probably a result of varied concentrations of odor-causing substances in different species.

Milk can also serve as a vehicle for foreign substances that are soluble in fat and water, which can cause unpleasant flavors. These substances might originate from the cow's diet as well as the environment prior to and following the milking procedure. Multiple factors contribute to taste abnormalities in milk. Unconventional milks, which are atypical or non-commercial milks not intended for gathering or trade, may possess a noticeably salty taste as a result of an anomalous protein and mineral content. Alterations to the constituents of milk, including variations in its chemical, physical, and microbiological makeup, can also result in alterations to its taste. Milk can undergo negative changes in its sensory quality due to microbiological issues, such as bacterial contamination. Oxidized fat can also cause undesirable odors in milk, as oxygen affects the phospholipids in milk, resulting in flavors such as metallic, cardboard, fishy, and greasy.

5.7. THE PROCESS OF FLAVOR DEVELOPMENT IN DAIRY PRODUCTS

The dairy industry faces a significant problem with the development of unpleasant flavors in milk, cream, and dairy products. However, a regulated process called enzyme-catalyzed hydrolysis of milk fat is often employed to intentionally create desirable flavors in specific products.

These applications encompass the production of lipolyzed milk fat products to enhance butter-like characteristics, the formulation of flavors in various types of cheeses, and the formulation of flavors in milk chocolate. Individuals examine the literature regarding the correlation between taste and the breakdown of fats, which encompasses several enzymes such as microbial and tissue lipase or esterase. Comparative data about the specificity and activity of various enzyme preparations demonstrates a wide spectrum of flavor impacts. The evaluation process focuses on applications for commercial purposes, such as patents, that involve the use of lipolytic enzymes or their byproducts in various types of cheese, goods derived from lipolyzed milk fat, recipes for bakeries, and taste preparations. The authors of the study also consider potential future applications (Kumar et al., 2017).

The initial microorganisms are crucial for the development of cheese's distinctive flavor. The main source of taste in cultured milks and fresh cheeses is the transformation of lactose into lactic acid by lactic acid bacteria, along with the production of diacetyl and acetaldehyde. The initial bacteria present in mature cheeses rapidly perish, and the speed at which they disintegrate and release their enzymes into the system impacts the rate of production of free amino acids (Kumar et al., 2017). Rennet acts alone to generate large, medium, and small peptides. Without any other enzyme interactions, it can only yield measurable quantities of methionine, histidine, glycine, serine, and glutamic acid.

The flavor of fresh bovine milk is mostly attributed to a set of chemicals that fall into chemical classifications such as free fatty acids, sulfur, carbonyl, and alcohols. Specific volatile compounds associated with the taste of milk include 2-butanone, ethyl caproate, hexanal, nonanal, octanal, and pentanal. Although handling and processing can lead to various microbiological and chemical-induced alterations in the taste of milk, it is becoming more evident that there are additional underlying factors at play (Kumar et al., 2017). Despite the preference of many individuals to consume cold milk directly from the refrigerator, a recent study suggests that the temperature at which milk is eaten has minimal influence on flavor perception. This is due to the fact that parameters such as fat content and other compositional elements have a more significant impact on the aftertaste feeling.

Non-fat milk was believed to possess a greater intensity of sour, bitter, cooked, and fatty flavors, as well as a little more chalky and dull taste compared to whole milk. This is attributed to its larger amount of hexanal and lower amounts of benzaldehyde, ethyl caproate, heptanal, 2-heptanone, and nonanal.

The five taste components, namely sweet, salty, sour, bitter, and umami, collectively form the entirety of human flavor perception. A meal reaches a higher level of quality when each element is skillfully harmonized, both in terms of presentation and overall taste.

Acquiring knowledge of the basic principles for preparing each element is the initial stage in grasping the concept of how flavors are harmonized. It is important to note that the addition of salt to food serves a dual purpose: it not only imparts a salty taste, but also enhances or harmonizes other flavors present in the food (Kumar et al., 2017). The subsequent explicit instructions delineate how each constituent will influence the overall taste:

- **Sweetness:** Derived from many sources such as fruits, sugar, honey, or other substances, serves to counterbalance the tartness and bitterness of flavors. Additionally, it can be employed to mitigate the spiciness of a dish.
- **Saltiness:** Salt serves two essential roles in enhancing the taste of food. To begin with, it mitigates feelings of animosity. Furthermore, it enhances the taste profiles of the majority of the other components in the dish, particularly the element of sweetness. The delectable taste of salted caramel arises from the perfect balance between the sweetness of sugar and the savory touch of salt. Similarly, salt is commonly included in recipes that showcase tomatoes in order to amplify their inherent flavors (Cadwallader and Singh 2009).
- **Bitterness:** Although not often favored, bitterness plays a crucial role in achieving balance. Dark leafy greens, lager beer, and grapefruit all possess qualities that can counterbalance the sweetness or richness of a dish.
- **Tartness:** Citrus and vinegar are commonly associated with this taste sensation. Acidity is a remarkable method to enhance vitality and counterbalance sweetness and spiciness in a food.
- **Umami:** The umami flavor, which is naturally present in soy sauce, mushrooms, oysters, and various cheeses, can be challenging to articulate. An optimal approach to utilize umami is by harmonizing it with other flavors, particularly in a dish that appears to be well-balanced but is still missing something.

Heat treatment of milk proteins leads to Maillard reactions, resulting in the formation of variou□flavor compounds including Strecker aldehydes, sulfur and nitrogen-containing compounds, maltol, and diacetyl (Cadwallader & Singh, 2009).

CHAPTER 6

DAIRY MICROBIOLOGY

Contents

It is widely recognized that bacteria that appear at various stages of production significantly affect the quality of cheese and other fermented foods. Due to the fact that different types of microorganisms have different metabolic abilities, it is not surprising that the study of dairy microbiology has focused on analyzing the makeup of cheese microbiota and how it contributes to achieving specific sensory characteristics (Robinson, 2005).

In recent decades, there has been a notable change in the investigation of the microbial ecology of intricate natural habitats. The availability and utilization of novel techniques for microbial detection, identification, and typing have led to this development. Advancements in molecular techniques that analyze DNA and RNA have opened up new possibilities in the field of environmental microbiology (Robinson, 2005). As a result of fascinating findings and progress in environmental microbiology, other branches of microbiology, such as food microbiology, have updated their research methods and improved their analytical tools. The scientific techniques and methodologies employed to study microorganisms in food have undergone significant transformations compared to previous eras.

The primary focus of dairy microbiologists is to study the diversity and changes in microorganisms throughout dairy production. They also investigate how specific microbial species and strains can influence the quality of cheese. The conventional culture-based approaches remain the standard and predominant approach for studying the microbiology of cheese. However, an escalating number of laboratories currently possess the essential instruments and technological expertise to examine the microorganisms present in cheese utilizing molecular techniques (Robinson, 2005). Traditional microbiology relies on the growth of microbes on appropriate, and sometimes specific, culture media.

6.1. MICROORGANISMS IN MILK

Microorganisms have a substantial impact on the quality and safety of milk. Although milk is initially sterile when it is produced in the udder, it can be contaminated with different microbes during the process of milking, handling, storing, and other pre-processing phases. Microorganisms can have either a beneficial or detrimental effect, and their existence can influence the deterioration and safety of milk and dairy products.

Lactic acid bacteria are a prevalent category of microorganisms that can be detected in milk. These bacteria metabolize lactose through fermentation to generate lactic acid and are commonly employed as starting cultures in the manufacturing of cultured dairy products such as yogurt. Lactic acid bacteria commonly present in milk include lactococci and lactobacilli. Coliforms, a type of bacteria that can survive with or without oxygen, are capable of contaminating milk. Although not always detrimental, coliforms serve as markers of possible pathogen presence and have the ability to deteriorate the quality of milk.

The microbial makeup of milk can serve as a means to evaluate the production conditions and sanitary standards of the product. The proliferation of bacteria in milk can lead to deterioration, while the presence of pathogenic microorganisms can taint milk and pose a potential hazard to consumers. Hence, it is crucial to implement preventive measures in order to minimize the probability of contamination and eradicate any microorganisms that can infiltrate. Conversely, some microorganisms have a positive impact on the synthesis of dairy products by inducing desired chemical alterations.

The primary cause of milk spoilage is psychrotrophic microbes, which are capable of thriving at refrigerator temperatures and are resistant to pasteurization. These organisms possess the ability to generate heat-resistant extracellular enzymes that degrade proteins, carbohydrates, and lipids, resulting in spoiling. Two bacterium species that can lead to milk spoiling are Pseudomonas fluorescens and *Pseudomonas fragi*.

Although hygienic milk production practices and pasteurization have significantly decreased the likelihood of milk borne diseases, there are still bacterial pathogens that may be found in raw milk or dairy products that have been inadequately pasteurized or mishandled. The pathogens mentioned are Bacillus cereus, Listeria monocytogenes, Yersinia enterocolitica, Escherichia coli O157:H7, *Salmonella* spp, and *Campylobacter jejuni*. Consuming milk that is contaminated with certain microorganisms might result in foodborne diseases.

Brucella species provide a risk in milk, as they can induce flu-like symptoms including fever, perspiration, migraines, lower back discomfort, and debilitation. *Salmonella jejuni*, a bacterium typically found in water supplies, poultry, and the digestive system of cattle, can also be detected in milk and lead to diarrheal disease. *Coxiella burnetii*, a heat-resistant pathogen, can be excreted in the milk, urine, and feces of sheep, goats, and cattle.

Implementing appropriate hygiene standards throughout the whole process of milk production, handling, and processing is crucial for ensuring the safety and quality of milk and dairy products. Pasteurization is an essential process for eradicating detrimental bacteria; however it may not always be fully successful against all categories of pathogens. Hence, it is imperative to uphold impeccable hygiene protocols across the whole milk production and processing cycle in order to mitigate the potential for contamination and guarantee the safety of milk and dairy products for consumers.

Ultimately, bacteria have a substantial impact on the quality and safety of milk. Although certain microbes can lead to food spoiling and constitute a potential danger to customers, others have a positive impact and play a role in the creation of appealing dairy products. Adhering to proper hygiene practices and implementing pasteurization are crucial for guaranteeing the safety and excellence of milk and dairy products. However, it is imperative to consistently monitor and control microorganisms at every stage of production and processing to minimize the chances of contamination and ensure the safety of these products for consumers.

6.2. MICROORGANISMS THAT CAUSE SPOILAGE

Yeasts, molds, and bacteria are the three primary classifications of microorganisms responsible for food spoilage. Bacteria, including those that cause food spoilage, are present and can lead to rapid and considerable deterioration of food. Yeasts and molds have a slower development rate compared to bacteria, but they can still flourish on different substances and diets. They also have a greater ability to withstand environmental difficulties (Wang et al., 2017). Food spoilage bacteria originate from several sources such as foods, their surroundings, processing equipment, and the food itself. Typically, the types of bacteria that cause deterioration in raw materials and finished commodities tend to vary.

6.2.1. Bacteria

Lactic acid-producing bacteria (LAB) such as *Carnobacteria*, *Weissella*, *Enterococcus*, *Lactococcus*, and *Lactobacillus*. Specific strains of LAB are utilized in the process of food fermentation, whilst certain other strains might contribute to the degradation of food. Undesirable modifications induced by LAB encompass the generation of gas, unpalatable flavors, the development of a viscous texture, discoloration, and the formation of slime on meat, among other effects (Wang et al., 2017).

Acetic acid bacteria (AAB) are prevalent microorganisms found in environments rich in ethanol and sugar. They often convert alcohol into acetic acid during the wine-making process, resulting in the deterioration of the wine.

Enterobacteriaceae have the ability to transform amino acids into sulfuric chemicals and smelly diamines. Spore forming microorganisms, specifically species of Bacillus and Clostridium, have the potential to induce degradation in various food items, particularly those that are canned or vacuum-sealed.

Yeasts commonly cause the deterioration of goods that have a high sugar, low organic acid, and high salt content. Consequently, these conditions commonly lead to the deterioration of food, particularly fruits, fruit-based goods, syrups, jams, sauces, alcoholic drinks, and confectionery. Common spoilage molds, such as *Penicillium and Aspergillus* can cause the appearance of mold. Some of these molds have the ability to produce mycotoxins. Fruits, grains, and cereals are commonly associated with fungal decay (Wang et al., 2017). The spoiling leads to the development of prominently pigmented growth, the formation of gas or acid, and the presence of unpleasant odors.

6.2.2. Microorganisms Controlled by Antimicrobial Packaging

Microbiologists are aware that foodborne pathogenic species may be accurately identified, but the microorganisms responsible for food spoiling are often the same ones that facilitate food fermentations, such as lactic acid bacteria, yeasts, and molds. Food spoilage bacteria are responsible for the breakdown of an item's organoleptic features, which are the properties of food that can be perceived by the human senses. These microbes affect these traits through their activity (Wang et al., 2017). Hence, the primary focus of endeavors to manage the proliferation of food spoilage bacteria is usually directed towards perishable food items such as meat, fish, milk, fruit, and fresh produce.

However, antimicrobial packaging can effectively be used in any case where natural factors (such as pH and water activity) are not enough to control the growth of spoilage bacteria. Many publications focus on the creation of antimicrobial packaging that can effectively combat rotting germs. A study has been conducted to evaluate the effectiveness of antimicrobial packaging using in vitro or challenge tests, similar to those used for infections.

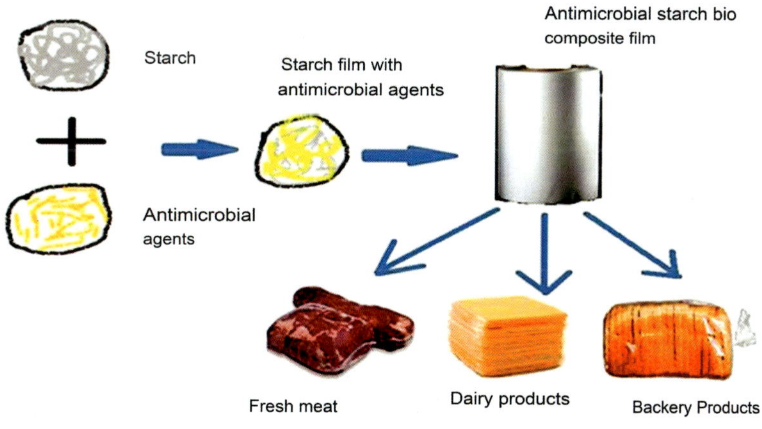

Figure 6.1. *Applications of antimicrobial starch biocomposite film.*

Source: Mallick et al., 2020.

The efficacy of this packaging, however, might also be assessed in research focused on controlling the local spoilage bacteria. The microorganisms that are typically evaluated include *Penicillium spp.* for fungi, *Brochothrix thermosphacta* for bacteria, *Carnobacterium* spp. for bacterium, *Lactobacillus plantarum* for plantarum, and *Pseudomonas* spp. for bacteria (Wang et al., 2017). Furthermore, different groups of microorganisms, including the total number of live microorganisms, enterobacteria, and lactic acid bacteria, are sometimes seen to verify the overall effectiveness of the antimicrobial package being studied (Figure 6.1).

Bacteriocins have been employed to develop antimicrobial packaging that effectively targets gram-positive food spoilage bacteria, owing to their antibacterial properties. However, the combination of other elements improves and expands their antibacterial properties. Utilized nisin, EDTA, and HCl to activate bags made of a coextruded barrier film (VDC/EVA/LDPE). After applying a solution containing a mixture of all antimicrobials to the inside surface of the LDPE, boneless beef chunks were sealed in vacuum-packed bags (de Blackburn, 2006). The samples clearly showed that using antimicrobial packaging had a significant effect on reducing the levels of rotting populations after 32 days at 1°C. However, the use of molecular technology indicated that this influence had minimal impact on species diversity.

The dairy industry is heavily reliant on spoilage bacteria, as they metabolize milk components to continue their activity and survive inside milk. Consequently, these bacteria cause significant changes to dairy products (de Blackburn, 2006). The entire eradication of spoilage bacteria from raw milk is unattainable due to their presence in the equipment, processing environment, and milking area. The milkers are primarily responsible for executing procedures to prevent contamination and hence reduce financial losses resulting from product alterations. Several bacterial species have the ability to spoil milk. The main alterations involve the breakdown of lipids and proteins caused by bacterial enzymes (lipase and protease) and the increase in acidity caused by saccharolytic microorganisms.

Consequently, they generate lactic acid as their main metabolic product, leading to a decrease in the pH of the milk. The decrease in pH affects the stability of milk by altering the charge distribution of casein around the colloidal micelle, which is normally negatively charged when the milk's pH is within the range of around 6.6–7.8 (de Blackburn, 2006). The decrease in repulsive forces between the micelles happens when the concentration of H+ ions in the solution increases, reaching the pH value of 4.6, which is the isoelectric point of casein. Moreover, the elevated concentration of H+ ions leads to the denaturation of the colloidal phosphate bonds, which is a vital process for preserving the link between the submicelles.

The mesophilic group of microorganisms, such as *Lactobacillus*, Streptococcus, *Lactococcus*, and some enterobacteria, is accountable for the process of acidification. Psychrotrophs are extensively found in the environment and, if they infect milk, they proliferate rapidly in low temperatures. This group includes species from the genera Pseudomonas, Bacillus, *Lactobacillus*, and Listeria (de Blackburn, 2006). Although most psychrotrophic bacteria are susceptible to heat treatments commonly employed in the dairy industry due to their thermolability, certain genera such as Bacillus, Clostridium, and Streptococcus exhibit thermoduric characteristics that allow them to tolerate the heat treatment. The most notable characteristic of psychrotrophs is their ability to manufacture heat-resistant lipolytic and proteolytic enzymes that maintain their enzymatic activity even after being subjected to heat treatment. These enzymes are capable of functioning throughout the entire shelf life of the milk and have specific regions where they operate on different components of the milk.

6.3. PATHOGENIC MICROORGANISMS

The World Health Organization (WHO) identifies bacteria, viruses, toxins, parasites, and chemicals as potential hazards that might be found in food. A pathogenic organism is an entity capable of inducing diseases in a host (human).

Pathogenic bacteria, such as Streptococcus, Pseudomonas, Shigella, Campylobacter, and Salmonella, can also affect other important diseases like pneumonia and foodborne infections (Torkar & Teger, 2006).

Pathogenic bacteria can be transmitted to humans through the consumption of milk and milk products. Pathogenic bacteria such as *Enteropathogenic Escherichia coli, Yersinia enterocolitica, Campylobacter jejuni,* and *Listeria monocytogenes* have recently become a focus of concern in relation to milk, cheese, and ice cream. Dairy products are globally manufactured and consumed as a component of a well-balanced diet. However, similar to other types of food, the presence of bacterial pathogens in food can be a risk to public health and can occasionally lead to outbreaks of sickness or disease (Torkar & Teger, 2006). Understanding the ecology of bacterial pathogens in the dairy environment is crucial for implementing effective food safety standards, as many of these pathogens are naturally found in dairy products.

The effective control of pathogenic microorganisms in the dairy chain remains a significant challenge in the manufacture of wholesome dairy products. Ensuring the production of high-quality, secure milk is essential for maintaining the safety of dairy products. The introduction of processing controls, such as heat treatment, has led to a substantial improvement in dairy product safety. The initial application of pasteurization was to enhance the stability of food. However, it was soon recognized that pasteurization could also be employed to deactivate microorganisms in liquid milk. In modern commercial pasteurization processes, foodborne bacteria are commonly rendered inactive through the use of turbulent flow at a minimum temperature of 72°C for a length of 15 seconds. Alternatively, a combination of temperature and duration that achieves similar results can also be employed (Torkar & Teger, 2006). The efficacy of this processing technique in deactivating significant bacterial pathogens, including *Listeria monocytogenes*, *Shiga-toxigenic Escherichia coli, Salmonella species, Cronobacter species,* and *Staphylococcus aureus*, is illustrated by kinetic modeling. Food additives, such as salt or nisin, have been introduced to prevent the growth or neutralize dangerous bacteria, and this intervention has had a significant impact (Torkar & Teger, 2006). Nevertheless, the domain that plays a crucial and ever-changing role in ensuring the safety of dairy food is likely the implementation of optimal standards for maintaining cleanliness on farms and throughout processing. The combination of these initiatives has contributed to the successful implementation of stringent safety requirements for dairy products in most developed countries. Understanding the environmental sources of these viruses is crucial for informing sanitary protocols aimed at preventing their introduction into dairy production environments and, ultimately, dairy products. A multitude of studies have been done to understand the ecology of dairy products and the routes by which they become polluted in the dairy chain. Given the prevalence of these pathogens in dairy environments, there is a possibility that milk may be contaminated by cross-contamination. Raw milk may become contaminated either during the milking process or if the bulk milk tank is not adequately sterilized. Hence, it is crucial to uphold proper hygiene in milking procedures to prevent the entrance of contaminants. Transporting tainted milk to the processing area creates a route for germs to infiltrate and can lead to cross-contamination of dairy products. Once within the processing area, these harmful microorganisms can establish themselves and lead to ongoing contamination of dairy products, posing a challenge for dairy food processors to eliminate these stubborn strains (Jezierska-Tys et al., 2010).

Presently, there are bacterial infections that are highly significant in terms of ensuring the safety of dairy food.

Listeria monocytogenes is a type of bacterium that can cause foodborne illnesses. It is classified as Gram-positive, meaning it retains a purple stain when tested using the Gram staining method. Additionally, it is a facultative anaerobe, which means it can survive and grow in both the presence and absence of oxygen. The genus is classified as containing 17 species, although only one of them, *L. monocytogenes,* is pathogenic to humans. Food regulation typically focuses on setting microbiological standards specifically for the presence of *L. monocytogenes* in dairy products. This is because pasteurization effectively deactivates *L. monocytogenes* in milk (Jezierska-Tys et al., 2010).

Shiga-toxigenic Escherichia coli (STEC) has either one or both of the Shiga toxin genes, stx1 or stx2. These germs are associated with debilitating complications such as renal disease and irritable bowel syndrome, as well as the severe condition known as hemolytic uremic syndrome, or HUS. Staphylococcal food poisoning (SFP) (*Staphylococcus aureus*) occurs when enterotoxins produced by the coccoid Gram-positive bacterium *S. aureus* are ingested through contaminated meals (Jezierska-Tys et al., 2010). Under favorable conditions, such as when dairy products are heated wrongly, the bacterium reproduces and typically creates these enterotoxins in food. Upon consumption, these thermally stable enterotoxins induce sickness, such as vomiting and/or diarrhea.

6.4. MICROORGANISMS INVOLVED IN FERMENTATION

Enzymes function in the metabolic process of fermentation to chemically modify organic molecules. Fermentation is a term commonly used in food preparation to describe any process where microbes cause a desired change in a meal or beverage. However, in biochemistry, fermentation is explicitly defined as the process of extracting energy from carbohydrates without the presence of oxygen (Figure 6.2).

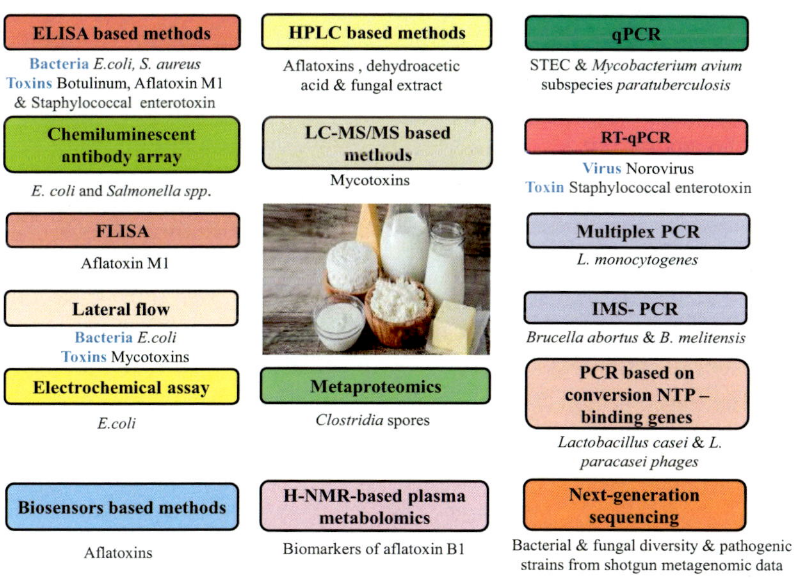

Figure 6.2. Different pathogenic microorganisms and toxins in fermented dairy products.

Source: (El-Sayed et al. (2022).

The primary genera in the lactic acid bacteria category are *Lactobacillus, Leuconostoc, Lactococcus, Pediococcus,* and *Streptococcus.* Microorganisms metabolize lactose, the primary form of carbohydrate found in milk, into lactic acid and various other byproducts (Fernández et al., 2015).

Throughout history, humans have ingested dairy products or milk that has undergone fermentation. Various strains yield distinct fermented products. Lactic acid bacteria are the predominant microorganisms responsible for performing lactic acid fermentation. These substances are categorized as GRAS (generally regarded as safe) because to their abundance in nature and presence in fermented foods, as well as on the mucosal surfaces of the human body. The main genera in the group of lactic acid bacteria are *Lactobacillus, Leuconostoc, Lactococcus, Pediococcus, and Streptococcus* (Fernández et al., 2015). These microbes metabolize lactose, the primary form of carbohydrate found in milk, into lactic acid and other byproducts. Due to the acid's effect on the milk's proteins, fermented goods generally have a denser texture compared to milk.

Lactic acid bacteria (LAB) species are widely utilized microorganisms in global food fermentation processes. The presence of LAB is crucial for the fermentation of milk, since it facilitates the transformation of milk into superior fermented milk products. Both naturally occurring and intentionally added starter cultures can contain LAB during the process of fermenting milk. These two cultures exhibit potential for further exploration in the manufacturing of fermented milk. LAB are essential in the fermentation process of milk, as they contribute to the production of acid. This acid not only enhances the flavor of the goods but also acts as a preservative. In addition, they produce exopolysaccharides, which are essential for the formation of texture.

The process of milk fermentation is a relatively simple and cost-effective technique. In contrast, standardized fermented milk products are produced using large-scale, regulated manufacturing methods, and they have become a key industrial application for LAB starter cultures. LAB starters possess several essential traits in fermented milk products. Unlike items that undergo spontaneous fermentation, the presence of a single starting culture in fermented milk products will reduce the variety of microorganisms present (Fernández et al., 2015).

6.4.1. Characteristics of Lactic Acid-Producing Bacteria

The role of LAB in the milk fermentation process is crucial, as they are necessary for converting raw milk into fermented milk products. Various industrial strains of LAB are used as starter cultures in the milk fermentation industry. The LAB starter cultures were obtained by following a sequence of steps, which involved isolating, selecting, and confirming them (Fernández et al., 2015). A wide range of behaviors and characteristics of each distinct strain of LAB have been identified and are utilized in the industrial production of fermented milk products. The most significant qualities of LAB are their proteolytic activity, which allows them to transform milk protein and contribute to the acidification, flavor, and texture of milk.

The prevalent bacteria involved in fermentation include Bacillus, *Lactobacillus, Micrococcus, Streptococcus, Pediococcus, and Leuconostoc* species. Moreover, the typical fungal species identified in fermented food items include *Cladosporium, Aspergillus, Trichothecium, Paecilomyces, Fusarium, and Penicillium.* Saccharomyces is the predominant yeast species used for the synthesis of alcohols (Jay et al., 2005). The development of fermenting microbes in fermented foods and beverages is influenced by characteristics such as pH, water activity, salt content, temperature, and food matrix composition. The process of food fermentation is facilitated by the most common microorganisms through the action of lactic acid bacteria (L.A.B.). This type of lactic acid fermentation provides enhanced nutritional value, improved safety, prolonged shelf life, and general approval.

Fermented dairy products such as sour milk, yogurt, acidophilus milk, and other similar milk products are in increased demand due to their superior nutritional value compared to conventional milk. Fermented milk has a variety of proteins, vitamins, carbohydrates, and varied types and quantities of fat, while retaining its mineral content. The quality of sour milk is determined by the chemicals produced during the biochemical processes of milk souring and the fermenting microorganisms involved. The constituents include alcohol, lactic acid, vitamins, carbon dioxide, and antibiotics (Jay et al., 2005).

Fermented milk is generated when milk is allowed to undergo fermentation by lactic acid bacteria, such as Lactobacilli. Some individuals find it easier to process or understand due to this factor. Fermentation breaks down proteins and lactose found in milk. Individuals who suffer from lactose intolerance or have an allergy to milk proteins will see advantages from this. In addition, fermented milk products have a longer shelf life. Fermented milk is employed for the management of lactose intolerance, hypertension, hypercholesterolemia, dermatitis, antibiotic-induced diarrhea, and common colds (Jay et al., 2005). Furthermore, it is purported to alleviate hay fever, asthma, constipation, stomach ulcers, and various other conditions; yet, most of these applications lack substantial scientific evidence.

Microorganisms can derive advantages from the rich nutritional content of milk, just as humans and newborn mammals. Raw milk at room temperature is prone to microbial decay. After a few days, the milk will naturally become sour. The activity of lactic acid bacteria is usually responsible for this. These bacteria can develop into a microbiota that can be intentionally added to fresh milk to maintain or potentially improve it. This concept is the basis for the meticulous control of milk acidity to create secure and durable goods with typically enjoyable tastes. Across the globe, a multitude of fermented milks and their derivatives have been developed, each with its own unique historical background (Jay et al., 2005).

The properties of the product are significantly impacted by the type of milk utilized, the pre-treatment of the milk, the temperature and conditions during fermentation, and the subsequent technical treatments applied. Lactic acid bacteria predominate in fermented dairy products, however other bacteria, yeasts, and molds may also be found. Efforts have been made to optimize the manufacturing processes of fermented dairy products by producing starter cultures (Jay et al., 2005). They consist of distinct microorganisms that are cultivated either as mixed-strain starters, comprising a random combination of various bacterial kinds, or as multiple-strain starters, composed of a precise blend of pure cultures.

Although the utilization of starters has significantly enhanced the quality of dairy goods, it has concurrently limited the assortment of fermented dairy products accessible. There is currently a resurgence of interest in searching for potential initial microorganisms from the pool that existed during the fermentation of raw milk. This is because the dairy industry is keen to explore new opportunities for broadening the range of products it offers.

6.5. MICROBIAL QUALITY CONTROL IN DAIRY INDUSTRY

Milk quality control is the systematic examination of milk and milk products to ensure their safety and healthiness. This process also verifies compliance with specific standards regarding purity, chemical composition, and the presence of bacteria and other microorganisms. Ensuring high milk quality is essential for consumer health and safety.

In order to achieve optimal outcomes from testing procedures, it is just a matter of applying logical reasoning and practical knowledge to quality control measures in the laboratory. Primarily, the responsibility resides with the manufacturer of the supplies and apparatus used, and it is upheld through their implementation in the dairy control lab.

Various microbiological count procedures, including the total bacterial count (TBC), coliform count (CC), and coagulase-positive staphylococci count (CPS), can be employed to evaluate the hygienic quality of milk (Ahmed & Abdellatif, 2013). The determination of the microbiological quality and safety of raw milk and the factors that influence it at the farm level is crucial, as it directly impacts the quality and safety of subsequent products during the manufacturing process.

While milk is highly nutritious and an essential component of human meals, it also serves as an ideal medium for the proliferation of germs, posing a risk to consumers' health. Raw milk, which is obtained from healthy cows' udders, contains a low concentration of bacteria, making it generally considered safe for human consumption (Ahmed & Abdellatif, 2013). The spoilage bacteria and pathogens that may contaminate milk after it is expelled from the udder can originate from various sources, including animal excrement, soil, air, feed, water, bedding material, animal skin, infected udder, the external surface of the udder and teats, milk containers, and human contact.

The quality of milk produced on a farm is determined by the procedures used in farm management. It is crucial to ensure clean milk production on the farm in order to prevent microbiological contamination of milk and to avoid health risks for customers. The quality of products across the dairy supply chain is influenced by the initial microbial load of raw milk at the farm level. For example, if the initial amount of germs is insignificant, additional heat treatments may not be required to entirely eliminate them. The main factors contributing to the presence of harmful microorganisms in bulk milk at the farm are: bacteria originating from the external surfaces of the udder and teats, particles from the milking equipment's surface, and mastitis-causing organisms from within the udder (Ahmed & Abdellatif, 2013).

The Total Bacterial Count (TBC) is the predominant method for evaluating the sanitary condition of unpasteurized milk on the farm, providing an approximation of the overall quantity of aerobic bacteria in the milk. TBC serves as a useful indicator of the level of cleanliness in the cow, the milking equipment, and the farm surroundings. The presence of contaminants in the environment, unclean udders and teats, lack of hygiene in milk equipment, and the presence of mastitis-causing organisms can all lead to an increase in the Total Bacterial Count (TBC) of bulk tanks (Ahmed & Abdellatif 2013). Storage temperature and duration since milking are other crucial factors in influencing the quality of milk, as they directly impact the rate of bacterial proliferation.

The coliform count is used to estimate the quantity of coliform bacteria present in milk. The environment in which the cow is located is the primary source of coliforms found in milk. Coliform bacteria, commonly originating from unclean udders and teats, indicate the presence of fecal matter in milk. High levels of colony-forming units per milliliter (> 10^3 CFU/mL) in raw milk may indicate unsanitary conditions, unhygienic milking practices and subsequent handling, contaminated milk equipment, polluted water, inadequate refrigeration, or the presence of coliform mastitis (Biegańska, 2022; Cadwallader & Singh, 2009;

Canning, 2020). A high CC value indicates the potential presence of pathogenic bacteria, such as *E. coli* O157:H7, which generates the Shiga toxin, in milk, posing a risk to public health (Granados et al., 2018).

The primary cause of bovine mastitis is predominantly *Staphylococcus aureus* (Capper, Cady, & Bauman, 2009; Chandan, 2015), resulting in substantial economic losses due to reduced milk yield and quality. In Ethiopia, the isolation of *S. aureus* is commonly attributed to bulk milk and cows affected by mastitis. The presence of *S. aureus* in bulk milk samples indicates the existence of infected cows in a herd. Due to the widespread consumption of raw milk in Ethiopia, □taphylococcal food poi□oning pre□ent□a □ignificant potential threat to public health (Granados et al., 2018). *Staphylococcus aureus* secretes enterotoxins that trigger food illness when ingested in contaminated food. Milk and milk products are frequently linked to staphylococcal food poisoning.

6.5.1. Ways to Control the Microbiological Quality

These tasks encompass the collection of samples, enumeration and identification of microorganisms, analysis and assessment of data, and ongoing surveillance of environmental conditions. Given the requirement for regular utilization, the equipment might be classified as an environmental consideration.

Only milk of exceptional quality can be utilized to produce dairy products of the utmost excellence. Evaluating milk is essential to guarantee that only premium-quality milk is sent to merchants and processors, while samples of lower quality are rejected. In order to do this, it is essential for farmer cooperatives and milk dealers to possess the knowledge of how to conduct milk testing on the milk they acquire from individual farmers or produce themselves (Granados et al., 2018). Dairy personnel must possess knowledge of various quality control protocols. They should possess the ability to identify and understand the underlying causes of odd tastes.

Sensory analysis: This test is the simplest, as it solely relies on the senses of sight and scent. If milk has a strong, unpleasant odor, an unusual color, or visible particles, it should be thrown out.

- **Coagulation Boiling (COB):** This is a brief and uncomplicated examination. It allows you to ascertain whether the milk has a high amount of acidity (pH below 5.8). Avoid milk with elevated acidity levels. The test described by Granados et al. (2018) can be used to differentiate between mastitic milk and colostral milk, which is produced in the early days after giving birth. Due to the high concentration of whey proteins in colostrum milk, it is advisable to avoid cooking or boiling it during processing to prevent any complications.

6.6. THE USE OF PROBIOTICS IN DAIRY PRODUCTS

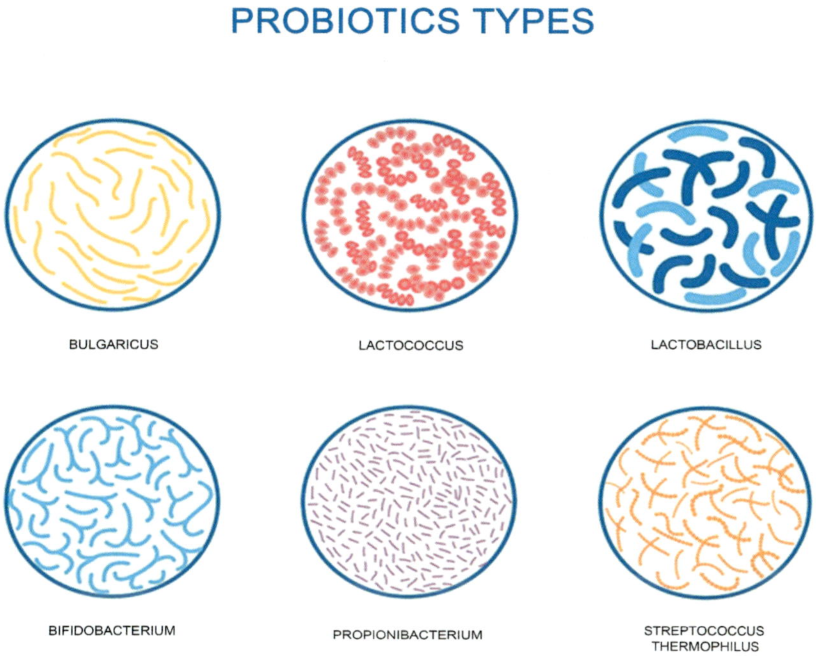

Figure 6.3. *Types of probiotics in dairy products.*

Source: Mejeriteknisk Selskab (2021).

Functional foods incorporate a substantial amount of probiotics. However, there has been limited discussion regarding its potential application in food fortification. Consequently, scientists have examined the possible products that can be generated when probiotics are present and active, as well as how their nutritional value may increase. Gao et al. (2021) conducted a recent study on the role of probiotics in converting minerals into amino acids, synthesizing B-complex vitamins and peptides. The study also explored new technologies and methods for evaluating these alterations (Figure 6.3).

Food is essential for maintaining good health as it provides energy, acts as a reservoir of nutrients, and ensures the functioning of bodily functions. Probiotics are a category of microorganisms commonly present in food, primarily recognized for their ability to enhance gastrointestinal well-being. Dairy products such as milk, cheese, yogurt, and even ice cream may contain these germs. Probiotic bacteria can flourish in dairy products due to the provision of optimal conditions such as temperature, pH, and time, as well as the availability of a suitable substrate with the necessary chemical composition. Moreover, these meals serve as a protective barrier for probiotics as they pass through the digestive system, enabling them to regulate colonization and provide extra bioactive components that enhance probiotic effectiveness (Gao et al., 2021).

Probiotics exert their effects on human health by successfully surmounting the biological obstacles, like as pH levels and digestion enzymes, present in the gastrointestinal tract. Some of the biological activities include antimicrobial, antimutagenic, antigenotoxic, lactose tolerance-enhancing, cancer prevention, and immunomodulation. Dairy products often include probiotics derived from the genera *Lactobacillus* and Bifidobacterium. The interaction between microorganisms and the matrix depends on whether they were introduced into the product during fermentation (Gao et al., 2021). Given that their occurrence can take place either prior to or following the cooling process, the number of interactions would be reduced if they were introduced at the end.

For the fermentation process to occur, the active participation of bacteria is essential. Therefore, it is important to consider aspects such as the chemical composition of the media and the composition of the starters. Probiotics and starter cultures exhibit antagonistic activity, which can impede the growth of bacterial components or even inhibit them. To prevent this, it is necessary to closely monitor the formation of lactic acid and the drop of pH during the fermentation process. Bacteria can obtain metals, vitamins, and peptides through their biosynthesis, making them a valuable resource for food fortification. Unfortunately, several countries still do not meet the recommended daily consumption of essential micronutrients.

Minerals are essential minerals derived from food and can be found in both organic and inorganic compounds. However, organic metabolism is considered more favorable. (Gao et al., 2021) Soil is the primary reservoir of minerals, which it also provides to plants and animals. Humans obtain essential minerals through the consumption of plants and animals. Minerals are found in biological fluids, tissues, and a significant percentage of the skeleton. These substances are crucial for maintaining human health, and although they must be consumed daily, just a small quantity is necessary to fully benefit from them. The recommended daily intake for secondary minerals is often between 400 and 1500 mg, whereas for micro minerals it is between 45 and 11 mg (Umer, 2014).

Dairy products include several minerals in the form of ions, salts, and organic compounds. Calcium, along with other minerals, can be either artificially included into items or occur naturally within them. Fortification is the act of adding bioactive, non-nutritive substances to a product along with nutrition. In milk, minerals and trace elements are typically present as inorganic ions, forming complex compounds with peptides, carbohydrates, lipids, and small molecules.

Probiotics have demonstrated the ability to synthesize vitamin B9, and a proposed mechanism for this process has previously been identified. Dihydropteroate synthase facilitates a condensation reaction that creates a carbon-nitrogen link between the precursors DHPPP and pABA, which are essential for the synthesis process. Afterwards, the dihydrofolate synthase enzyme adds glutamate groups to the dihydropterin (DHP) molecule, forming dihydrofolate (DHF). The DHF is then converted to the active cofactor tetrahydrofolate (THF) through reduction, and it combines with polyglutamate to produce THF-polyglutamate (Umer, 2014).

The antihypertensive action of probiotics is attributed to the suppression of angiotensin-converting enzyme (ACE). While the activity of ACE has been hindered by cultures of *Lactobacillus* and *Lactococcus* bacteria, milk from different animal sources has been the most commonly used medium for its synthesis. Yogurt and Capri kefir have only been investigated once for their synthesis, as documented by Umer in 2014. Following the separation of peptide fractions from the milk samples, the IC50 value was determined in order to ascertain the protein concentration needed to inhibit 50% of the ACE activity.

6.7. SAFETY MEASURES IN DAIRY PROCESSING

Post-milking handling procedures should aim to maintain the nutritional integrity of the milk and prevent any degradation caused by biological and physical factors that may occur during milk processing. In order to guarantee the safety of milk, it is imperative to adhere to both industry and governmental rules when it comes to the maintenance of farm equipment.

As an employer and dairy farmer, it is incumbent upon the farmer to provide their employees with a secure work environment. At the workplace, it is necessary to ensure the well-being and security of volunteers, contractors, guests, and other individuals. Therefore, it is imperative that you maintain a hazard-free and disease-free environment in your workplace (Chandan, 2015). The safety management system should aid individuals in overseeing and monitoring the following domains to effectively manage safety inside and around farm dairies.

6.7.1. Employee Training and Onboarding Processes

Initiating an orientation is the primary step to take while recruiting fresh personnel for a dairy farm. During the induction process, it is important to cover a range of themes, including the property layout, employment expectations, and fundamental information. In addition to the fundamental induction, it is necessary to provide a health and safety induction (Chandan, 2015). The health and safety data should encompass your company's policies and processes, risks and dangers, emergency protocols, as well as instructions and training specific to their respective jobs.

6.7.2. Engagement with Contractors

Effective communication is crucial for implementing a comprehensive risk management strategy. While you have attended to the onboarding of your workers, it is important not to neglect your obligations towards your contractors. During certain seasons, it may be necessary to engage the services of contractors and subcontractors to assist with additional tasks such as silage or fencing. Both the contractor and the property owner are jointly responsible for ensuring the contractor's health and safety while on the property. Before the Contractor accesses your property, they should have knowledge of the specific locations of your effluent ponds and your established guidelines on the management of animals and physical work (Chandan, 2015).

6.7.3. Regulations and Directives

When it is required, furnish evidence to demonstrate that your regulations and protocols are establishing a secure working environment. To mitigate health and safety risks, it is imperative to establish guidelines that clearly outline the responsibilities of both management and personnel. Dairy farmers are required to adhere to many regulations, including those related to working in enclosed areas like feed silos and pits, transporting livestock, reporting accidents and dangers, and operating machinery and equipment.

6.7.4. Examinations and Confirmations

Identifying potential hazards prior to posing a threat to the personnel is an integral component of the risk management process. Instead of engaging the services of a safety expert, utilize the comprehensive range of inspections offered by the Safe Ag Systems suite to effectively detect and address the specific hazards associated with the dairy business. One can effectively manage animal welfare, biosecurity, safe and responsible animal treatment, and dangers in their cattle yards by utilizing a variety of dairy inspection templates. Equip personnel with a handheld device to enhance their awareness of potential hazards (Chandan, 2015). Generate tasks for the purpose of monitoring safety inspections, or utilize your laptop or mobile device to review and analyze the inspection report.

The quality of raw milk in dairy processing is contingent upon various aspects, such as animal health and cleanliness, feed and water quality, milking procedures and equipment, and storage and transit conditions. It is essential to conduct tests to assess the physical, chemical, and microbiological properties of raw milk, including temperature, acidity, fat content, protein content, somatic cell count, and bacterial load, in order to ensure its quality. Moreover, it is advisable to refuse or separate raw milk that fails to meet quality criteria or shows signs of adulteration, contamination, or spoilage (Ntuli et al., 2023). Finally, it is crucial to adequately refrigerate and preserve raw milk to prevent the growth of microorganisms and maintain its quality.

Pasteurization and heat treatment are essential techniques for reducing the number of harmful microorganisms and enzymes in dairy products and extending their shelf life. The achievement of this can solely be attained by subjecting the product to a specific temperature for a defined duration, followed by a rapid cooling process. To ensure efficient pasteurization and heat treatment, it is necessary to monitor and regulate the temperature, pressure, and length of the heating and cooling phases using sensors, timers, and indicators (Ntuli et al., 2023).

CHAPTER 7

QUALITY CONTROL IN THE DAIRY INDUSTRY

Contents

Milk quality control refers to the systematic examination of milk and milk products to ensure its safety, healthiness, and compliance with specific standards regarding purity, chemical composition, and the presence of bacteria and other microorganisms.

Quality management is crucial in the dairy sector since it has a direct impact on consumer health, safety, and satisfaction. It is important to not only adhere to regulations, but also to ensure that dairy products are maintained at the highest possible quality, promote brand recognition, and utilize technology to monitor the quality of dairy products (Mostert & Jooste, 2002). Quality management is the unwavering dedication to achieving the highest standards of quality that is present throughout the whole dairy supply chain, spanning from the farm where the milk is produced to the table where it is consumed.

Dairy quality management is responsible for ensuring that dairy milk maintains a consistently high level of quality. It refers to being highly conscientious about following sanitary protocols and safety precautions during the processes of shipping and milking (Mostert & Jooste, 2002). Rigorous testing and monitoring protocols are implemented to detect any variations in milk quality, ranging from bacterial contamination to fat composition. By implementing these strategies, dairy companies may ensure the integrity of their dairy milk and the excellence of their final products.

The dairy business adheres to an intricate regulatory structure that aims to safeguard consumers and guarantee the excellence and safety of dairy products. Regulatory standards encompass a wide range of areas, such as labeling, food safety, and hygiene. Compliance with these regulations is legally obligatory for dairy operations and must not be ignored. Effective dairy quality management is crucial for maintaining adherence to regulatory standards for dairy products. The task involves developing software applications for quality management systems and other solutions that enable dairy enterprises to monitor and record their compliance with regulatory regulations. These solutions enhance operational efficiency, minimize the risk of regulatory violations, and result in time and cost savings by automating and optimizing compliance procedures (Mostert & Jooste, 2002).

7.1. CHEMICAL ANALYSIS OF DAIRY PRODUCTS

Every dairy product is a complex mixture of many chemical compounds. These chemical species are commonly categorized into key groups, such as water, fat, protein, lactose, and minerals, for the purpose of proximate study. If all chemical species in their natural states were classified into a single category, then the total of each separate category should add up to 100%. However, this is seldom the situation in practice. The presence of both random and systematic errors in the testing process, along with occasional specific assumptions about the chemical properties of a class (such as assuming the nitrogen content of dairy proteins to be 15.67%), contribute to this phenomenon (Lee et al., 2015).

Chemical analysis is conducted for a multitude of reasons:

- Label veracity, or the capacity to authenticate the information provided on the label. For example, the additional calcium found in skim milk that has been fortified with calcium, or the additional vitamins and minerals that are added to milk powders specifically designed for use as infant meals.
- Ensuring access to an adequate and reliable food supply. The levels of nitrates and nitrites, which are artificial compounds that form in milk products during processing, need to be lower than the specified values.
- Compliance with the Food Regulations. For example, butter must contain a maximum of 16% moisture.
- Compensation to agricultural producers. The purpose of this is to ensure that farmers receive equitable and just compensation for their milk.
- Describing the standard composition of milk. This provides the Dairy Board with a basis to compensate the industry for specific products created using a standard milk composition.
- Assessing the key components of the milk upon its arrival at the facility. Due to the significant impact of milk composition on product processing, it is necessary to employ efficient and accurate measurement methodologies.
- To provide the necessary data for the purpose of genetically improving the national herd. The fat and protein levels in cows differ according to the breed, as well as the genotypes within the same breed. Milk scans, which are rapid automated instruments, are utilized for conducting many standard chemical tests (Lee et al., 2015). These equipment are adjusted to measure protein, fat, lactose, moisture, and total solids according to established procedures.

The dairy industry employs a range of chemical techniques, ranging from basic to complex, and from classical wet chemistry to modern instrumental methods.

Chemical analysis include use concentrated sulfuric acid to identify formalin, iodine solution to detect starch, and nitric acid testing to identify skim milk powder in milk samples.

Chemical analysis encompasses several techniques, such as employing concentrated sulfuric acid to identify formalin, utilizing an iodine solution to detect starch, and Chemical analysis, which relies on measurements, is divided into two sections based on the methods used for conducting the assays. Wet chemical analysis, often known as classical analysis, pertains to analytical techniques that just necessitate a balance and do not require any further mechanical or electrical apparatus (Lee et al., 2015). Usually, the method relies on chemical reactions that take place between the analyte, which is the substance being analyzed, and a reagent that is applied to it. Wet methods often depend on the production of a chemical reaction product that can be easily measured and identified.

Most chemical analysis is categorized as instrumental analysis, belonging to the second group. It involves conducting the analysis using a tool that is not a balance. The analyst has a wide array of instrumentation available to them. The device can be utilized to quantify a specific characteristic of a substance or to evaluate the nature of a chemical interaction between the substance and another agent. Instrumental analysis categories are established based on the kind of instruments utilized (Barbano & Lynch, 2006).

Analysis is conducted to ascertain the composition of a product. An analysis determines both the quality of the product and the quantity of ingredients required for production. Payment is contingent upon several crucial product attributes.

The composition of milk is determined by its SNF (Solids-Not-Fat) percentage and fat content. In the industry, the estimation of SNF (solids-not-fat) is done using a lactometer, while the estimation of fat content is done using the Gerber technique. The Electronic Milk Tester is employed in certain dairy farms to quantify the fat composition of milk (Barbano & Lynch, 2006). Only a small number of dairy farms utilize milk scans to ascertain the levels of SNF (solids-not-fat), fat, protein, and lactose in their milk. Conducting tests to determine acidity, detect adulterants, and identify preservatives is essential in order to evaluate the quality of a product.

Proficiency in testing, suitable laboratory practices, and method knowledge are necessary for the analysis of milk and milk products.

The freshness of milk is characterized by the presence of carbon dioxide, citric acid, albumin, casein, and minerals, which collectively result in its significantly low acidity. Subsequently, lactose acidity increases due to bacterial activity during storage. The acidity of milk is determined by titrating it with a standardized alkali solution and utilizing phenolphthalein as an indicator, as alkali has the ability to neutralize acid. The acidity of milk generally ranges from 0.14% to 0.16%. A concentration of less than 0.12% in the lower range signifies neutralization, but a concentration of more than 0.16% in the higher range suggests the occurrence of acidity due to bacterial activity or mastitis milk (Barbano & Lynch, 2006).

The term "Gerber Method" refers to a rapid technique used to estimate the fat content in liquid milk. The basis of this method involves quantifying the quantity of fat released from a specified volume of milk sample using a specially built and accurately calibrated glass cylinder called a butyrometer.

Most testing methods employ gross compositional analysis to determine the quantities of ash, protein, fat, moisture, lactose, and total solids. One of the other tests conducted is macro elemental analysis, which includes the examination of calcium, magnesium, sodium, potassium, phosphate, nitrate, and nitrite. Another part is dedicated to the analysis of trace elements, vitamins, and other naturally occurring minor constituents such as cholesterol and fatty acids. A study of the dairy sector has undergone significant development and transformation in the past decade, as evidenced by the work of Barbano and Lynch (2006).

7.2. EXAMINATION OF THE PHYSICAL PROPERTIES OF DAIRY PRODUCTS

The Dairy Technologist highly values the physical properties of milk and milk products due to their significant impact on various processing operations. These operations include fluid flow, mixing and churning, emulsification, homogenization, as well as heat transfer processes like pasteurization, sterilization, evaporation, dehydration, chilling, and freezing. Rheological properties are utilized to assess and monitor the quality of certain products, such as yoghurt, milk, butter, and cheese (De Beer, 2012).

An exemplary raw milk should possess clarity, devoid of any silt or debris, and should exhibit no peculiar flavors, weird colors, or odors. It should have a low germ count, be free from chemicals such as detergents and antibiotics, and maintain a consistent composition and pH level. The primary factor that influences the quality of milk products is the quality of the raw milk. Exclusive, high-quality raw milk is required to produce top-tier milk products.

It is crucial to implement proper hygiene procedures at every stage of the dairy process to ensure the milk's sanitary quality. This is necessary to produce milk and milk products that are safe and suitable for their intended use (De Beer, 2012). The challenges faced by small-scale dairy producers in producing hygienic products can be attributed to informal and unregulated marketing, handling, and processing of dairy products, a dearth of financial incentives for quality improvement, and a lack of knowledge and expertise in hygienic practices.

Basic milk testing methods suitable for small-scale dairy farmers and processors in developing countries consist of the Gerber test for quantifying milk fat content, the clot-on-boiling test for detecting sourness or abnormalities, the density meter or lactometer test for measuring lactic acid levels, and the organoleptic tests involving taste, smell, and visual examination (De Beer, 2012).

Milk possesses various fundamental physical attributes, such as density, conductivity, thermal properties, rheological behavior, redox properties, colligative properties, surface activity, buffering capacity, and color.

The PCC Genetic Improvement Program receives help from the Dairy Laboratory, which offers milk component analysis services. The main target demographic for this service consists of dairy herds that actively engage in performance testing and milk recording, particularly cooperatives that receive help from PCC. Lactose, total solids, somatic cell count (SCC), milk fat percentage, and milk protein percentage can be rapidly assessed using milk samples. The PCC National Headquarters presently has the facility accessible (De Beer, 2012).

Typically, the analysis is completed within nine business days from the date of the request or the submission of the milk sample. By using bronopol, raw milk samples can be preserved for a duration of seven days at a temperature range of 4–5°C. If the milk sample is not preserved (without the addition of bronopol), it should be placed in a container packed with ice and promptly sent to the laboratory for analysis as soon as possible (within two days) (De Beer, 2012). Unpreserved samples can be analyzed within 72 hours after the first collection of the milk sample. Prior to sending samples for orientation or training on proper milk sampling, handling, and storage procedures, potential clients should first coordinate with the milk laboratory personnel.

Standardized analytical techniques are essential in the trade of secure, high-quality dairy products because they provide a basis for shared understanding, assess compliance with local, state, and federal regulations, and serve as a benchmark for existing techniques or as an introduction for routine inspection and control (Bintsis et al., 2008). Due to the voluntary collaboration of various stakeholders in the dairy sector, such as the industry itself, government agencies, researchers, and manufacturers, they all use the same vocabulary, understanding, and standards to ensure consistent and comparable outcomes.

The text discusses the techniques for collecting milk samples and the various tests available to measure the quantities of fat, solids, protein, lactose, acidity, and other constituents present in milk (Bintsis et al., 2008). These tests include freezing point determination, butterfat content analysis, lactometer testing, alcohol testing, and organoleptic testing, among others.

An ideal raw milk should be free from any silt or debris, exhibit no unusual flavors, possess a typical composition and pH, have minimal microbes, and be devoid of any additions such as detergents or antibiotics. The raw milk's quality is the main determinant of the quality of milk products. Only high-quality raw milk (Bintsis et al., 2008) is allowed to be used in the production of milk-based products. Every stage of the dairy supply chain should undergo rigorous quality control and testing procedures for their milk. It is feasible to test for the following contaminants in milk:

- Quantity: indicated by either weight or volume.
- Organoleptic features: encompass visual characteristics, taste, and scent, as well as compositional attributes such as protein, fat, and solid content.
- Physical-chemical properties: Some examples of hygienic characteristics include the presence of medication residues and adulteration with substances such as water, preservatives, or extra solids.
- Sanitary conditions, hygiene, and high-quality standards.
- Organoleptic tests, such as taste, smell, and visual examination, are suitable methods for evaluating milk for small-scale dairy producers and processors in developing countries.
- The exact density of milk can be determined with a lactometer or a density meter.
- The clot-on-boiling test is employed to detect abnormal or sour milk.
- Evaluating the acidity of milk to determine its lactic acid concentration.
- The fat content in milk is determined by the Gerber test.

7.3. MICROBIOLOGICAL ANALYSIS OF DAIRY PRODUCTS

Antibiotics are commonly employed in cattle production to promote growth and prevent diseases. Animal products such as meat and milk may include residues as a result of improper usage or excessive dosage. These residues pose a worldwide health danger when they are consumed and enter the human body, as they can lead to antibiotic resistance (Laslo & György, 2018).

Microbiological testing safeguards this microbial habitat. It provides dairy farmers with a crucial tool to ensure the safety and quality of their products by:

Identifying and quantifying the existence of microorganisms: This encompasses both beneficial bacteria that contribute to taste and consistency, as well as harmful pathogens such as *E. coli* and Salmonella. Microbiological testing aids in safeguarding people from the hazards of foodborne illness by detecting and eliminating harmful germs.

Preserving the quality of dairy products is crucial as spoilage bacteria can have detrimental effects on their flavor, texture, and shelf life (Laslo & György 2018). Microbial testing aids in the detection and control of organisms, ensuring that consumers may access fresh and high-quality products.

Certain beneficial bacteria, such as probiotics, have a positive impact on human health. Microbial testing enables the verification of the presence and appropriate quantities of these microorganisms in dairy products. Diverse approaches are employed to detect and identify microorganisms that are present in these items. Commonly employed methodologies encompass traditional culture-based techniques, molecular biology methodologies, expedited approaches, and targeted evaluations for mandatory and optimal evaluations.

Conventional culture-based techniques entail introducing a diluted dairy product sample onto an agar plate containing certain nutrients that facilitate the proliferation of bacteria. individual strains of bacteria can be selectively targeted by utilizing different varieties of agar that cater to their individual development needs. After incubation, colonies are then investigated and identified based on visual observation, size measurement, and other distinguishing traits. This approach is characterized by its ease of use and affordability. However, it may require a significant amount of time and may not be entirely effective in identifying all microorganisms, particularly those that have a slow growth rate or are experiencing stress.

The Most Probable Number (MPN) approach utilizes a series of tubes containing broth and varying dilutions of dairy samples. The tubes are placed in an incubator, and the presence of any bacterial growth is seen and recorded. The MPN is calculated through statistical analysis of the number of tubes that yield positive results. This procedure is more intricate than conventional plating and necessitates supplementary resources and expertise.

Molecular biology techniques, such as polymerase chain reaction (PCR), are employed to detect and amplify the DNA sequences of certain bacteria for identification purposes. PCR is a versatile technique that can be employed to detect and distinguish many species, including detrimental pathogens and bacteria responsible for rotting. PCR necessitates the use of specialized instruments and expertise due to its great sensitivity and precision. Real-Time PCR is a sophisticated method that enables the immediate measurement of microorganisms, offering quick screening and monitoring abilities.

Genomic sequencing is a molecular biology technique used to ascertain the whole DNA sequence of an organism. This data can be utilized for accurate discernment and categorization of bacterial variations, aiding in the determination of contamination origins and comprehension of the assortment of bacteria found in dairy products.

Rapid techniques, such as adenosine triphosphate (ATP) bioluminescence, quantify the amount of ATP, a substance found in all living cells. There is a direct relationship between the high number of microbes and increased ATP levels. Although this technology has the ability to produce data quickly, it lacks the capability to differentiate between various categories of microorganisms. Immunoassays employ antibodies to precisely detect and identify antigens that are found on the exteriors of bacteria.

They yield data in a matter of minutes or hours and demonstrate a high level of sensitivity and specificity. Flow cytometry is a technique that examines individual cells by utilizing lasers and fluorescent dyes to assess their size, shape, and other distinguishing features. It has the capability to quickly detect and measure certain microorganisms.

The investigation of dairy products from a microbiological perspective can be divided into two main categories: mandatory examinations and optional assessments. Mandatory exams prioritize the identification and measurement of disease-causing bacteria, such as Salmonella, *E. coli*, Listeria monocytogenes, and *Staphylococcus aureus*. These germs present substantial health hazards, and regulatory authorities determine and oversee permissible limits for particular illnesses. Indicator microorganisms, such as coliforms and Enterobacteriaceae, are often examined to evaluate the general hygiene of the dairy processing facility. Their existence suggests inadequate hygiene or pollution throughout the production procedure.

Bacterial count is an essential test that measures the total quantity of bacteria, irrespective of their specific species, found in a dairy product. This assay offers insights into the durability of the product and its level of hygiene. The maximum allowable numbers for bacterial count can vary depending on the specific product and applicable legislation.

Effective evaluations prioritize the identification of harmful microorganisms that cause spoiling, such as Pseudomonas, Bacillus, and lactic acid bacteria. An overabundance of these microorganisms can result in alterations in texture, disagreeable flavors, and diminished longevity. Depending on the kind of dairy product and the dangers associated with certain bacteria, it may be necessary to test for specific spoilage organisms (SSOs). For instance, fresh milk can undergo testing to detect psychotropic bacteria, which have the ability to thrive at cold temperatures and lead to spoiling during storage.

7.4. SENSORY ANALYSIS OF DAIRY PRODUCTS

The most reliable measure of a successful and superior product is its sensory characteristics. Sensory analysis employs a diverse array of powerful and delicate tools to measure individuals' responses to various foods and products. Identifying the appropriate test, configuring the appropriate parameters, and evaluating the data yield consistent, meaningful, and prompt results. By doing these tests correctly, one may interpret the study of volatile chemicals in relation to taste perception, as well as get valuable insights on unique products and consumer preferences (Drake, 2007).

When talking about dairy products, there is always a certain level of sensory quality involved. The sensory quality of dry dairy components and fluid milk is of utmost importance, as the highest quality raw ingredients result in the best final products. It is crucial to additionally take into account the tactile sensation of completed products such as cheese and ice cream. The enduring appeal of dairy products can be attributed, in part, to their sensory perception, which is both delicious and beneficial to health (Drake, 2007). In several tests or applications, a sensory measurement is the final stage due to the essential role of dairy products in sensory perception. Occasionally, a rough measurement is sufficient to evaluate the quality or consistency of a product.

The enduring importance of sensory quality is undeniable, and individuals are inherently driven by capitalism to promote and sell the most exquisite and recently produced commodities due to their high demand. Similar to other scientific fields, sensory science has undergone continuous development and is currently in a state of evolution. Therefore, it is essential to employ the most advanced instruments that align with the objectives of the study in its current form. In order to accurately measure or evaluate human reactions to stimuli, several scientific approaches have been developed that are precise, replicable, and unbiased. The field of sensational analysis offers insights into a wide range of topics beyond mere preferences and aversions (Drake, 2007). Furthermore, one can also consider the impact of emotional responses and how these influence the way consumers perceive a product or service.

Oftentimes, sensory testing is conducted retrospectively, leading to the utilization of inaccurate or inadequately selected tests. When a scientific test is chosen or carried out improperly, it is impossible to acquire reliable data and make appropriate findings.

7.4.1. Different Types of Sensory Tests

Sensory analysis can be categorized into three primary classes or test categories. The initial set of examinations employs traditional instruments such as ADSA scorecard evaluation and USDA grading. The dairy business developed these sensory tests in the early 1900s to ensure the quality and consistency of its products, as well as to inspire and educate pupils. These approaches assign a grade or overall quality score to a product based on a pre-established list of issues (Drake, 2007). Originally employed in product research, these instruments now serve primarily for training students, resolving product quality issues, and ensuring the minimum quality standards of government commodities, due to the advent of modern sensory testing methods.

These strategies are highly efficient and effective in large production plants or quality control settings where rapid decisions on product quality are necessary. A proficient individual can promptly identify significant deficiencies in the quality of a product, together with their potential causes, and implement suitable repair or preventive measures. Nevertheless, scorecard grading and scoring systems are deemed unsuitable for market or product research due to certain methodological deficiencies (Lim et al., 2022). Scoring and identification are subjective due to the lack of clear and up-to-date definitions for the terminology used to characterize quality and grade problems. The quality/grade issues do not provide a complete description of the products.

7.4.2. Standard Sensory Assessments

- *Analytical Sensory Tests*: Analytical testing and affective or consumer tests are the two primary classifications of mainstream sensory tests. These tests are well-established and scientifically recognized, distinguishing them from commodity-based traditional sensory tests. Each category has specific sets of sensory tools designed for particular objectives, and each set can include a diverse range of individual tests or techniques. The discrimination or difference test is a widely recognized analytical sensory examination. The primary objective of a difference test, in its simplest and most widely used format, is to determine whether there are any distinctions between two or more products (Lim et al., 2022).
- *Examinations at the Point of Transition:* The threshold test is an analytical sensory test specifically designed to discover thresholds. A threshold refers to the minimum concentration at which a sensory reaction becomes noticeable.
- *Descriptive Analysis:* This analysis represents the final overarching classification of analytical sensory testing. Descriptive analysis involves training a small group of individuals (often six to twelve) to identify and quantify specific sensory attributes of food, including all of its sensory qualities (Lim et al., 2022). The panelists for this sensory tool must undergo training, unlike the previous analytical tests that involved individuals who were either informed or vetted.

The level of intricacy in the sensory attributes being analyzed dictates the extent of the required training.

7.5. REGULATORY STANDARDS FOR DAIRY PRODUCTS

In countries like Kenya and India, the regulatory standards for dairy products are defined by the KS1552:2016 Code of sanitary practice for milk and milk products and the guidelines set by the Food Safety and Standards Authority of India (FSSAI) respectively. The purpose of these regulations is to guarantee the safety and quality of milk and its byproducts at every stage of production, handling, and distribution.

Kenya has a set of specific requirements that must be met for dairy products. It is crucial to ensure that milk is promptly delivered to a collection center within a two-hour timeframe from the time of milking. For optimal preservation, it is recommended to keep it at a temperature of 10°C or lower for an extended period. It is a requirement for all milk delivery vehicles to prominently display the name and address of the licensed distributor or transporter. It is imperative that any equipment, utensils, and instruments that come into contact with milk are constructed from food-grade and approved materials, ensuring they can be easily cleaned and disinfected. Direct selling of raw milk to customers is prohibited in municipalities. Compliance with these regulations is overseen by the Kenya Dairy Board and local government. It is important for farmers to go through the necessary registration and licensing process. In certain rural and semi-rural areas, there may be allowances for direct sales of milk, taking into consideration factors such as the time since milking and the distance from the point of production (Coleman, 2001).

In India, the FSSAI has implemented guidelines to guarantee the safety and quality of milk and milk products. The guidelines encompass a wide range of topics, including processing, packaging, distribution, cleanliness of dairy farms, and the quality of cattle feed. The primary objective of the FSSAI is to ensure the well-being of consumers and enhance their trust in milk and its byproducts. According to the guidelines, it is important to note that the term "Milk" specifically pertains to the secretion of a healthy milk animal after it has been fully milked, without any additions or extractions. It is important to note that the milk must be free from colostrum, as mentioned by Coleman in 2001.

Regulatory bodies in both Kenya and India have the important task of overseeing and enforcing compliance with these standards. The Kenya Dairy Board is responsible for ensuring strict adherence to safety regulations and maintaining high dairy standards in Kenya. One of their responsibilities is to ensure the quality and safety of milk and milk products. They also conduct inspections and issue permits to dairy business owners, all in accordance with the regulations set forth in the Dairy Industry Act. Inspections are carried out with the aim of ensuring compliance, and licenses may be suspended if noncompliance is found. Surveillance is conducted on a regular basis to ensure strict compliance with safety and quality standards (Gandhi et al., 2020).

In India, the FSSAI plays a crucial role in setting and maintaining the standards for milk and milk products. Inspectors are responsible for conducting thorough examinations of milk handling facilities to ensure strict adherence to regulations and other crucial public health laws. Obtaining licenses from the FSSAI is a mandatory requirement for operators of dairy businesses. To ensure compliance, regular checks are conducted. The FSSAI conducts regular surveillance throughout the dairy value chain, offering laboratory testing and analysis, and assisting in decision-making concerning the safety and quality of dairy products. Collaboration with government and industry organizations is crucial for enhancing service delivery, as highlighted by Gandhi et al. (2020).

In both Kenya and India, the regulatory standards for dairy products have been put in place with the primary goal of guaranteeing the safety and quality of milk and its byproducts. The regulations encompass a wide range of areas including production, handling, and distribution. They are put in place by regulatory bodies to ensure the well-being of consumers and foster trust in dairy products.

7.6. TRACEABILITY IN DAIRY INDUSTRY

Traceability is the ability to monitor and track the movement of goods or resources across the supply chain. Traceability is crucial for the dairy industry's endeavors to uphold food safety, ensure quality assurance, and adhere to regulatory requirements.

The dairy product supply chain is at the forefront of a transformative shift, driven by the dual requirements of sustainability and traceability. The dairy industry is undergoing transformation in order to meet the evolving demands of customers who seek accountability and openness in the food they consume (De Las Morenas et al., 2014). Stakeholders are implementing innovative approaches to enhance traceability and promote sustainability throughout the whole supply chain, spanning from the source to the end product. Organic dairy products, including milk, cheese, and yogurt, are experiencing increasing popularity in the market due to a growing focus on personal health and well-being. Traceability has been a significant concern in the food industry since consumers want to know more about the safety and quality of the food they select (De Las Morenas et al., 2014).

Traceability is required for the dairy business due to the following factors:

- **Ensuring Regulatory Compliance:** The dairy business is subject to multiple regulations and stringent quality criteria. In order to ensure compliance with these stringent regulations, traceability is not merely a choice, but an absolute need. A robust traceability system is crucial in the entire milk supply chain, encompassing procurement, processing, and distribution, to showcase accountability and unwavering adherence to the intricate regulatory framework.
- **Protecting Quality:** Temperature, handling, and storage conditions are just a few of the environmental elements that dairy products are extremely susceptible to. The traceability system is a precise instrument that enables you to promptly identify the origin of any deviations in quality and implement suitable measures (De Las Morenas et al., 2014). This ensures the consistent quality of your products and safeguards the reputation of your organization.
- **Maximizing Inventory Control Efficiency:** Effective inventory management is crucial for preventing both excessive stocks and the risk of severe supply shortages. Through the implementation of traceability, you can accurately and effectively track and monitor the levels of inventory in real-time. This helps to minimize wastage and ensures that the appropriate quantity of stock is maintained.
- **Using Advanced Data Insights:** The dairy sector is not an exception to the rule that data is important in this day and age. Utilizing a cutting-edge dairy supply platform provides you with advanced data analytics capabilities that thoroughly examine your operational environment (De Las Morenas et al., 2014). Analyzing this data enables one to discern emerging patterns, uncover potential chances for cost reduction, and pinpoint areas that are primed for enhancement.
- **Transparency in the Transportation of Raw Milk:** The raw milk source is a crucial component in the dairy supply chain. Through the use of traceability, you may effectively monitor and assess the reliability and efficiency of your milk carriers. This ensures the integrity of your production process by ensuring the timely and secure provision of raw milk to your processing facilities.
- **Building Supply Chain Resilience:** Unforeseen interruptions in the supply chain might arise at any moment in today's contemporary setting. An robust foundation for the dairy supply chain enhances your ability to promptly adapt to fluctuating conditions. This platform provides your supply chain with the ability to effectively

manage unexpected events, such as adverse weather conditions, transportation issues, or other unforeseen circumstances (Mania et al., 2018).

The dairy supply chain has several issues that affect its efficiency, sustainability, and resilience. These concerns encompass the need to trace milk batches from various farms, which can pose difficulties for small-scale farmers lacking advanced monitoring tools. In the event of contamination or quality problems, it is essential to have robust procedures in place to accurately determine the milk's origin.

Various factors, including disease outbreaks, natural disasters, labor shortages, and transportation problems, can disrupt the movement of milk from farms to processing facilities and retail outlets. As a result, this can lead to shortages in supply and difficulties in distribution (Mania et al., 2018).

Ensuring adherence to food safety and quality standards is essential in the dairy supply chain to prevent contamination, spoilage, and the occurrence of foodborne illnesses. Ensuring the safety and purity of dairy products is crucial throughout the entire process, starting from the farm and ending at the table. This is achieved by rigorous sanitation techniques, quality assurance systems, and adherence to regulatory standards.

The dairy sector is receiving more scrutiny for its environmental impacts. The impacts encompass concerns regarding the release of greenhouse gases, the consumption of water, and the deterioration of land associated with the cultivation and production of dairy products (Mania et al., 2018). The long-term sustainability of the dairy supply chain relies on implementing sustainable practices such as reducing carbon emissions, enhancing water efficiency, and promoting biodiversity. Furthermore, these policies contribute to the mitigation of environmental hazards.

7.7. ENHANCING EFFICIENCY IN DAIRY PROCESSING

According to estimates, global milk production has grown by 61% over the past 25 years, with an average annual rise of 2%. It is projected that milk production will reach 177 million metric tons by 2025 (Smit, 2003). The production of milk and dairy products continues to increase, indicating strong consumer loyalty towards dairy products as a whole, despite the emergence of plant-based milk and dairy substitutes. Despite the adoption of advanced manufacturing ideas such as robotics, lean manufacturing, and industry, dairy farms remain predominantly reliant on manual labor. Nevertheless, each significant stage in the transformation of raw milk into value-added products inside a dairy processing facility necessitates a substantial amount of machinery once it is provided.

The dairy sustainability framework (DSF) was established by the global dairy sector in October 2013. The purpose of this cooperative and non-competitive initiative is to assess and make known the continuous sustainability achievements of the dairy sector to the general public (Smit, 2003). Currently, over 30% of DSF members are responsible for global milk production. These members have made a commitment to actively adopt sustainability initiatives and measure their progress. The membership count of both existing and growing dairy economies continues to increase. The primary sustainability criteria used by DSF encompass the environmental, social, and economic aspects of sustainability, thereby ensuring alignment with the worldwide dairy sector. Fostering active participation from members is a vital element of achieving alignment. DSF encourages members to engage in discussions regarding the opportunities, challenges, and achievements they have experienced in the field of sustainability (Smit, 2003).

Enhancing the production of cow milk is imperative for dairy farms to enhance their profitability. The number of cows that can be grown is constrained by the restricted availability of stalls and feed. In order to enhance milk output at the farm level, it is imperative to optimize the productivity of each individual cow. However, it is essential to conduct a thorough and comprehensive audit of the herd before implementing an appropriate remedy.

The key determinants that can contribute to an increase in a farm's profit margin through milk output are as follows:

- The primary lever for maximizing revenue: This is the sales volume, which refers to the quantity of milk sold. Increased production volume and commercialization provide a decrease in the burden of escalating fixed costs such as heating and energy (Smit, 2003).
- The sales price: This is mostly influenced by the farm's potential and the current market conditions. Indeed, there exist specific strategies at the farm level that can be employed to enhance profitability.
- Another approach entails regulating the quantity of fat and protein present in milk, as well as its overall composition.

In addition to nutrition, the quantity of milk produced per cow on a daily basis is directly impacted by the conditions in which they are raised and the way the herd is managed. Prior to proceeding, it is imperative to carry out an audit. The outcome of the audit will be valuable in identifying areas that require enhancements (Smit, 2003). This approach will consider various factors, including diet, health, the environment, and housing conditions for herds, with a particular focus on the following aspects: Managing issues related to reproduction: Inferior reproduction quality extends the average duration of nursing by increasing the gap between calvings. Impaired reproductive function can directly impede the optimal utilization of food (Moró Pérez, 2013).

Health problem management: The various diseases commonly observed in dairy herds can hinder the efficiency of digestion (such as ketosis, acidosis, fatty liver, alkalosis, etc.), result in abnormal feeding behaviors (such as acidosis, lameness), and harm secretory tissues (such as mastitis, etc.).

Improving living circumstances is crucial for promoting an animal's behavioral potential. Animals require continuous access to the feed bunk and drinking water for their well-being. Preventing the occurrence of diseases that are peculiar to certain raising conditions requires suitable housing that takes into account factors like as ventilation, lighting, bedding quality, and stall arrangement.

Enhancing milk production from cows is crucial for dairy farms to enhance their profitability. The number of cows that can be grown is restricted by the limited availability of stalls and feed. Increasing milk output at the farm level requires maximizing the productivity of each cow.

The five S pillars, namely sort, set in order, shine, standardize, and sustain, offer a systematic strategy to effectively plan, organize, clean, create, and sustain a productive work environment (Moró Pérez, 2013).

CHAPTER 8

PACKAGING AND STORAGE OF DAIRY PRODUCTS

Contents

Milk, being a fundamental food, contains essential elements. Therefore, preservation solutions for dairy products must effectively maintain the integrity of their vitamins and minerals, as well as their flavor, consistency, and scent. The packaging of dairy products plays a crucial role in this context. The challenges encountered by enterprises involved in the packaging of dairy products vary depending on the specific type of product, such as whole milk, milk-based products, butter products, or cheese (Rejeesh & Anto, 2023) (Figure 8.1).

Figure 8.1. *Dairy product packaging.*

Source: Tiwari (2019).

Lactic acid-resistant, liquid-tight packaging is necessary for yogurt drinks, and milk beverages. They are unable to harm the microorganisms inside. Their ability to operate as a barrier is limited only by their shelf life. Yogurt is typically packaged in portion cups that have a thick structure and are sealed with caps made of plastic, aluminum foil, or paper coated with metal. To prevent contamination, the dairy products are stored separately from the processed metal due to the presence of a lacquer coating on the underside of the lid.

The packaging for milk and yogurt drinks should be durable and designed in a way that allows for easy sipping. The utilization of plastic bottles with a screw cap and lid film, or a three-layer composite carton. Ice cream packaging must possess the ability to resist tearing and puncturing in order to be suitable for deep freezing. Tube bags made of paper or plastic film are used for popsicles, while trays made of plastic or coated cardboard with sealing film and a resealable lid are used for portion and family packs (Rejeesh & Anto, 2023).

8.1. PACKAGING MATERIALS FOR DAIRY PRODUCTS

Plastic milk containers are utilized for the transportation, storage, and distribution of milk. Plastic bottles, generally referred to as jugs, have largely supplanted glass bottles for domestic purposes. Lightweight plastic bottles are designed for single-use and plastic recycling, while glass milk bottles are intended for several uses.

The packaging of milk is governed by regional authorities. It is vital to utilize items that come into contact with food; it is strictly prohibited to contaminate food. It is necessary to strictly follow processing and cleanliness norms. The main component of milk packaging is high density polyethylene (HDPE), which is categorized as recyclable. Polyester (PET) and low-density polyethylene (LDPE) are also used. Despite considering the use of polycarbonate, concerns arose over potential contamination with bisphenol (Ghenai, 2012).

Blow-molded plastic milk bottles have been in use since the 1960s. The primary material used is HDPE, with occasional utilization of polyester. Milk bottles come in a wide variety of forms. While certain objects have a square or rectangular shape, others possess a circular cross section. A novel square milk jug with a level top was recently developed to enhance storage and transportation efficiency; nevertheless, dispensing posed difficulties. Milk bottles frequently incorporate handles into their design. Additional milk bags are currently in use. When purchasing milk, it is often packaged in a plastic bag and then transferred into a pitcher. Occasionally, larger bags are employed for dispensing in institutions, serving as the inner bladder of a Bag-in-box. Small-sized individual milk and cream containers are commonly manufactured using injection molding or thermoforming techniques, and typically come with a detachable cap.

According to Ghenai (2012), pasteurized milk stored in HDPE bottles and LDPE pouches can last for a period of ten to twenty-one days when kept at a temperature of 4–8°C. Light and temperature abuse are further factors that exert influence. Aseptic processing and hyper pasteurization both extend the duration that a product can be stored without spoiling.

The quantity of milk in containers intended for retail sale must correspond to the quantity indicated on the label. Furthermore, it is imperative that the containers are visibly filled to ensure client satisfaction. Consequently, it is crucial to meticulously control the volume of the container. The precise capacity of the final blow-molded bottle is always uncertain to the die designer. After the item is removed from the mold, shrinkage occurs inevitably. Several factors, such as the duration of the cycle, the air pressure during inflation, the time the item is stored before filling, the temperature at which it is stored, and other variables, influence the extent of shrinkage (Ghenai, 2012).

An alternative method for making small changes to the capacity of a bottle without the need for building a new mold is to employ a volume adjustment insert. A circular cavity is created on the side of the finished bottle by affixing a volume insert to the interior of a mold. Inserts of different dimensions can be utilized in situations where production circumstances, such as mold temperature or cooling pace, undergo alterations. At regular intervals, the capacity of finished bottles is measured, and volume inserts are replaced as needed.

Currently, the most often utilized material for fresh yogurt is thermoformed TiO_2 colored high impact polystyrene (PS-HI). This material can be sealed with either a paper/plastic laminate or an aluminum foil/plastic laminate lid or closure. If a long shelf life is needed, barrier laminated materials are the ideal choice for pasteurized, spoon able yogurt products. Certain products in question have a shelf life of 4–6 months when stored at normal temperature. In order to prevent these items from losing moisture over time, it is essential to have a low water vapor transmission rate (WVTR) (Simon & Hansen, 2001). An effective light barrier will mitigate light-induced oxidation and delay the fading of light-sensitive hues. An effective oxygen barrier packing will aid in preserving the goods by preventing oxidation.

Drinking yogurt products are packaged in bottles made of PE-HD, which are sealed with caps made of PE-LD or heat-seal closures made of aluminum foil laminate. Additional varieties of plastic bottles, such as PET ones, can also be employed. Composite materials made of plastic, Alu-foil, or paperboard with excellent water vapor, oxygen, and light barrier properties are commonly used for durable, heat-treated drinking yogurt products. Yogurt is commonly enclosed in materials that are not transparent, such as opaque or semi-opaque substances. These materials often include a white pigment, such as TiO_2 (Simon & Hansen, 2001).

Modified atmosphere packaging (MAP) can be employed to prolong the shelf life of dairy products and effectively manage fungal problems associated with them. Dairy product packaging can utilize gas compositions that range from 10% to 100% carbon dioxide (CO_2), which is balanced with nitrogen (N_2) as an inert filler gas. This is done to prevent the package from collapsing due to the dissolution of CO_2 in the cheese product.

Metal, specifically aluminum, is used to make containers, foils, and closures. Laminates are commonly used to integrate foils with paper and plastic components in order to produce flexible packaging and food and beverage containers. Due to its robustness and stiffness, metal packaging allows for rapid filling without significant product wastage, resulting in enhanced overall efficiency, including reduced CO_2 emissions and energy consumption (Simon & Hansen, 2001). Due to its ease of handling in retail settings, metal packaging reduces spoilage and costs. Due to its impenetrable nature, the probability of leakage, tearing, or damage during transportation, storage, display on store shelves, or use in customers' homes is significantly reduced. Metal is utilized in several ways within the dairy sector, including the usage of aluminum foil for packaging butter, as well as cans for containing condensed milk and milk powders.

8.2. PACKAGING TECHNIQUES IN DAIRY INDUSTRY

Packaging is of utmost importance in the dairy business, particularly for products that have been subjected to thermal treatment. Heat-resistant packaging materials are essential for guaranteeing the quality and safety of liquid milk. Aluminum foil is commonly used for dairy products due to its ability to create a strong barrier against tastes, light, and oxygen. Paperboard is additionally employed to augment durability and rigidity.

The development of food packaging materials and technology is influenced by several variables, such as the desire of consumers for foods that have undergone little processing, modifications in retail and distribution methods, the introduction of new products, developing distribution patterns, and stringent regulations for consumer health and safety. The emergence of innovative packaging technologies, including modified atmosphere packaging, active packaging, intelligent packaging, and nano packaging, can be attributed to these considerations.

Modified atmospheric packaging (MAP) is a technique that includes removing the air from a package and replacing it with one or more specific gases. This process is used to extend the shelf life and improve the overall quality of food. This approach has been discovered to have a positive effect on extending the microbiological and sensory shelf life of cheese samples. There are several controlled atmosphere systems that can be used commercially, such as hypobaric storage, N-part generators, external gas generators, and gas apparatus systems.

Active packaging is a cutting-edge technology that includes active ingredients or technologies to improve the quality and safety of the packaged product. The process entails integrating adhesive substances into packaging film or containers to enhance food preservation, safety, sensory characteristics, and shelf life. For instance, the utilization of a film in packaging probiotic food has been discovered to effectively decrease both the presence of noxious odors and the concentration of dissolved chemical elements during the storage period.

UHT milk, which is packaged with a film that eliminates chemical components, similarly demonstrated a decrease in dissolved chemical elements over time while being stored. Furthermore, it was discovered that a laminate, which had been altered to include an iron-based chemical element absorbent, effectively decreased the levels of unidentified chemical elements in hot-filled paste (Patel et al., 2015).

To prevent oxidation and preserve the vitamin C content of beverages, it is advisable to keep them in PET containers equipped with a chemical element scavenger.

The use of oxygen absorbers or scavengers in packaging is an effective method to remove oxygen from the package. Iron is commonly used as a chemical element in packaging structures to chemically eliminate oxygen. The process involves the oxidation of iron and ferrous salts to generate a stable oxide. Various chemicals, including enzymes, sulfites, boron, and photosensitive dyes, can also be used as oxygen scavengers. Oxygen absorbers can be categorized based on their activation mechanism, scavenger form, and reaction speed.

Managing ethylene levels is crucial for prolonging the postharvest lifespan of fresh produce during storage. Ethylene acts as a growth-promoting hormone that accelerates the ripening process and reduces the shelf life of fruits and vegetables. Ethylene scavengers, such as potassium permanganate, can effectively absorb ethylene and convert it into carbon dioxide and water.

Antimicrobial packaging is another important aspect of food packaging. It effectively inhibits or halts the proliferation of harmful or decay-causing microorganisms by releasing antimicrobial substances. This can be achieved by applying bioactive agents to the surface of the packaging material, inserting a sachet containing an antibacterial component within the package, or dispersing bioactive agents throughout the packaging film. Antimicrobial packaging prolongs the lag phase of microorganisms and reduces the time they spend in the development phase (Patel et al., 2015).

The milk packaging industry is experiencing growth due to the increasing demand for efficient and innovative packaging solutions, particularly in developing countries. Consumer awareness of food safety and hygiene is driving this demand. However, the market expansion is hindered by stringent government regulations regarding packaging materials and environmental concerns. The demand for environmentally friendly packaging materials and the emergence of advanced technologies like smart packaging offer potential avenues for market expansion.

Several market trends are affecting the milk packaging industry. Sustainable packaging is becoming increasingly important as customers demand ecologically responsible packaging solutions. Smart packaging innovations, such as QR codes and freshness indicators, improve customer convenience and product traceability. Personalized packaging is also gaining popularity, with customers seeking unique products and the option to customize milk packaging. Collaborations between dairy producers and packaging manufacturers are driving innovation in milk packaging options (Patel et al., 2015).

8.3. SHELF LIFE OF DAIRY PRODUCTS

Understanding the shelf life of dairy products is crucial when considering their overall quality and safety for consumption. The shelf life of a food product is determined by the duration it can be stored without undergoing decay or becoming hazardous. There are various factors that can have an impact on the shelf life of dairy products. These factors include the type of packaging used, the conditions in which the products are stored, the techniques used during production, the quality of the raw materials, and whether or not the product has been opened.

Proper temperature control is of utmost importance when it comes to storing various types of food, including dairy products. It is essential to store dairy products, especially, in a refrigerated environment at all times. There are certain exceptions to consider when it comes to storage. For instance, powdered milk and UHT (ultra-high temperature) goods can be conveniently stored in the pantry. On the other hand, ice cream requires proper freezing to maintain its quality.

Dairy products stored in the refrigerator can have varying shelf lives. As an illustration, it is worth noting that pasteurized milk has a shelf life of 12–14 days when stored below 4°C. On the other hand, yogurt and other fermented products can maintain their quality for a longer period of 30–40 days under the same storage conditions. Various dairy products, including hard and semi-hard cheeses, cottage cheese, ricotta, cream cheese, salted butter, unsalted butter, ice cream, and frozen yogurt, each have their own specific shelf life durations.

Every dairy farmer must take on the crucial task of determining the shelf life of their products. During the specified period, it is essential for the product to maintain the required sensory, chemical, physical, and microbiological qualities. Several factors can impact the shelf life of dairy products. These factors include the quality of the raw materials used, the proper maintenance of the cold chain (ensuring milk and milk products are stored below 5°C), and the implementation of effective hygiene practices or other methods to ensure production safety.

Heat treatment is a widely employed processing technique in the dairy industry. The main purpose of this process is to eliminate any potentially harmful microorganisms and prevent spoilage, thus ensuring the safety of the products and prolonging their shelf life. Nevertheless, the duration for which a dairy product remains fresh is not solely dependent on its heat treatment. Additional factors, such as pH and organic acids, can also have an impact on the shelf life of fermented products.

In recent decades, the dairy industry has expanded its reach globally, with products being exported across borders. One important consideration is the adjustment of shelf-life specifications to accommodate various factors such as distribution, transportation, storage conditions, and customer preferences. Extending the shelf life of dairy products and shifting from refrigeration to room temperature storage can effectively minimize food waste and optimize the utilization of raw materials, ultimately improving the sustainability of the industry.

Technological advancements have played a significant role in the creation of dairy products that can now be stored for longer periods of time. Take, for instance, the remarkable advancements in ultra-high temperature (UHT) technology. Thanks to this innovation, we can now enjoy ambient spreadable processed cheeses. Additionally, the inactivation of culturing bacteria and the assurance of microbiological stability through low pH have paved the way for the production of ambient long-life yogurt. These innovations have a significant impact on the convenience and profitability of dairy products. Additionally, they create opportunities for new distribution channels, such as e-commerce.

Various factors can influence the shelf life of dairy products. These include packaging, storage conditions, production techniques, raw material quality, and whether or not the product has been opened. It is important to consider these factors when assessing the longevity of dairy products. Proper temperature control is of utmost importance when it comes to storing dairy products, as the majority of them necessitate refrigeration. Heat treatment is a widely employed method for

8.3.1. Factors Affecting Shelf Life of Dairy Products

- **Raw Materials:** The quality of the raw materials used in dairy products plays a significant role in determining their shelf life. Milk obtained from healthy cows that are free from diseases and infections will have a longer shelf life compared to milk obtained from unhealthy cows. Similarly, the quality of other ingredients used in dairy products, such as cheese cultures or yogurt starters, can also affect their shelf life (Figure 8.2).
- **Processing Techniques:** The processing techniques used in the production of dairy products can greatly impact their shelf life. For example, the pasteurization process, which involves heating the milk

to a specific temperature to kill harmful bacteria, can extend the shelf life of milk and other dairy products. Ultra-high temperature (UHT) processing, which involves heating the milk to a higher temperature for a shorter period of time, can further extend the shelf life of dairy products.

Figure 8.2. *Shelf-life of dairy products.*

Source: Garfield (2017).

- **Packaging Techniques:** The packaging of dairy products is crucial in maintaining their quality and extending their shelf life. Proper packaging helps to protect the products from external contaminants, such as bacteria or light, which can cause spoilage. Packaging materials that provide a barrier against oxygen and moisture can also help to prolong the shelf life of dairy products.
- **Storage Conditions:** The storage conditions in which dairy products are kept can significantly affect their shelf life. Dairy products should be stored at the appropriate temperature to prevent the growth of bacteria and other microorganisms. Refrigeration is commonly used to extend the shelf life of dairy products, as it helps to slow down the growth of bacteria. Additionally, dairy products should be stored away from direct sunlight, as exposure to light can cause the breakdown of certain components and lead to spoilage.

8.3.2. Mitigation and Control Strategies

To ensure the maximum shelf life of dairy products, several mitigation and control strategies can be implemented:

- **Quality Control:** Implementing strict quality control measures throughout the production process can help to ensure that only high-quality raw materials are used in the production of dairy products. This includes regular testing and monitoring of the raw materials for any signs of contamination or deterioration.
- **Good Manufacturing Practices (GMP):** Adhering to GMP guidelines can help to minimize the risk of contamination during the processing of dairy products. This includes maintaining clean and hygienic production facilities, training employees on proper hygiene practices, and regularly sanitizing equipment and utensils.
- **Proper Packaging:** Choosing the right packaging materials and techniques can help to protect dairy products from external contaminants and extend their shelf life. This includes using packaging materials that provide a barrier against oxygen and moisture, as well as implementing proper sealing techniques to prevent leakage.

8.4. STORAGE CONDITIONS FOR DAIRY PRODUCTS

Most dairy products require refrigeration, except for those prepared with powdered milk and UHT (ultra-high temperature) products, which can be stored in the pantry. Ice cream, on the other hand, must be kept in the freezer.

Due to the perishable nature of milk and most dairy products, they are prone to microbial contamination. Consequently, any deviations from the ideal temperature during storage and transit can adversely affect their quality and sensory attributes. It is crucial to be conscientious about this, especially during the summer. Utilize insulated bags when purchasing milk and milk products. It is recommended to promptly place recently purchased dairy products in the refrigerator and maintain a temperature between 4 and 6°C (Galstyan et al., 2019). If the products are kept in their original packaging and properly sealed, they will remain viable until the given date on the packaging, regardless of the circumstances. Milk and milk products undergo accelerated degradation after being opened due to the colonization of microbes from the surrounding air.

Acquiring dairy good□and milk: Prior to finalizing your purcha□e at the grocery store, ensure that you have added milk and milk products to your shopping cart. Choose milk from opaque jugs or cardboard cartons. Light may quickly and easily cause milk to become sour, which is why containers that allow light to pass through, such as translucent ones, are used.

Distributing dairy products: It is advisable to promptly bring dairy goods home from the grocery and immediately store them in the freezer or refrigerator. To enhance insulation and minimize cooling, it is advisable to either wrap frozen food around dairy products or employ insulated bags when transporting them throughout the summer season. According to Galstyan et al. (2019), specific perishable things can be safe for only thirty minutes in a heated car.

8.4.1. Methods for Utilizing and Preserving Dairy Products in a Domestic Setting

Immediately after purchasing, store milk and milk products (such as yogurt and cheese) in a refrigerator set at a temperature of 40° F or below. Retain milk in its original container. Ice cream and other dairy products should be promptly stored in the freezer at a temperature of 0° F or below. Milk should not be stored in the refrigerator door due to the exposure to fluctuating temperatures caused by frequent opening and closing of the door. After utilizing milk, instantly return it to the refrigerator. Do not return previously consumed milk to its initial container after it has been drained (Galstyan et al., 2019). Seal milk containers tightly to prevent the absorption of taste. The milk is safe, despite the potential for flavor alteration.

Irrespective of whether it is whole or skim, milk should always be stored in a refrigerator at a temperature of 37°F, usually in the back on a lower shelf. In order to maintain a comfortable fit, it may be necessary to store the milk horizontally. Avoid storing milk on any of the shelves located near the entryway. The convenience of easy access makes it an attractive location for storage. However, the front area tends to be warmer, and frequent opening and closing of the door cause temperature fluctuations that speed up the spoilage of milk (Galstyan et al., 2019).

8.4.2. Proper Milk Storage in the Freezer

To prolong the storage of milk, it is advisable to freeze it. However, it is crucial to use a plastic container for freezing, as milk expands during the freezing process. Cardboard and glass containers, being prone to breakage, are not suitable for this purpose (Loudiyi et al., 2022). It is possible to freeze milk prior to its expiration date, provided that there is enough space for it to expand, typically about an inch or two, and a plastic jug, such as the ones commonly used for milk, is suitable for this purpose. Defrost it in the refrigerator and let it defrost slowly. Proper storage of half-and-half and thick cream Like milk, cream, and half-and-half, they have optimal storage conditions in a cool

□ection of the refrigerator. Avoid approaching the door! It i□a door □pecifically designed for storing condiments. For extended preservation, cream can also be frozen in a plastic container or an ice cube tray. After defrosting, it is necessary to thoroughly mix or blend the substance to allow the liquid and milk fat to recombine. Thawed heavy cream can be whisked as effectively as fresh cream (Loudiyi et al., 2022). In addition, half-and-half can be frozen effectively using a freezer bag, plastic container, or ice cube tray.

What is the proper method of storing yogurt? It is advisable to refrigerate yogurt in it□natural □tate. Yogurt □hould be con□umed within a period of five to seven days after opening, however it may remain edible for up to two weeks. Discard the item if you observe pink streaks or mold, as the potential danger i□□ignificant (Loudiyi et al., 2022). Fortunately, yogurt may be preserved by freezing it when there i□an exce□□amount and in□ufficient time to con□ume it. Several prominent corporation□have generated □ub□tantial profit□by marketing frozen yogurt, a product that essentially consists of frozen granola infused with flavoring□ Yogurt i□mo□t enjoyable when kept frozen, however it doe□ experience a slight change in texture when thawed and refrozen.

8.5. TRANSPORTATION OF DAIRY PRODUCTS

People across the nation regularly consume dairy products. Many individuals simply pick up a carton of milk at the grocery store and proceed to leave. Many customers often overlook the process behind bringing a product to the market. Transporting dairy products requires a level of complexity that may not be immediately apparent.

Automation plays a crucial role in streamlining work processes, allowing farmers to increase productivity and enhance the quality of their products. Additionally, it provides the ability to constantly monitor the health of their herds, ensuring optimal conditions for their livestock. Additionally, it aids in the efficient tran□portation of dairy product□from farm□to con□umer□ table□ ensuring faster delivery and improved traceability. Technology has become an invaluable tool for dairy farmers, processors, and distributors, allowing them to streamline and automate tasks that were once carried out manually (Djekic et al., 2018). Personal cleanliness is an essential aspect of maintaining good hygiene and overall well-being. It is crucial to prioritize regular bathing, washing hands, and keeping our bodies clean to prevent the spread of germs and diseases. By practicing good personal cleanliness habits, we can promote a healthier and more hygienic environment for Considering hygiene is crucial when considering the secure transportation of dairy products. In the modern era, many people have concerns regarding their personal hygiene. It is essential for customers to ensure that the items they purchase are clean before taking them home. It is of utmost importance to ensure that the truck used for transporting dairy products is kept in a clean and sanitary condition. It is important to thoroughly clean the car both before and after each journey. Through this process, the products are ensured to arrive at their destination in a pristine condition.

Refrigeration is necessary for dairy products because of their sensitivity to temperature. It is important to keep dairy products properly chilled to prevent spoilage. It is crucial to maintain a cool temperature for dairy products at all times, including during transportation. Refrigerated trucks play a crucial role in the transportation of goods at specific temperatures (Djekic et al., 2018). Refrigerated vehicles play a crucial role in ensuring the safe transportation of dairy products to their final destination. In order to ensure the safety and freshness of the goods, these vehicles are equipped with insulation and refrigeration systems.

It is crucial for dairy products to be delivered on time to ensure they are consumed before they spoil (Djekic et al., 2018). Every dairy product container is marked with a printed date. The date stated on the packaging indicates the optimal quality of the product. It is important for customers to use the product before the specified date. In order to ensure timely delivery of the product, it is imperative that it be transported without delay, allowing customers to make their purchases before the deadline expires. Transporting dairy products becomes effortless and secure with the use of a refrigerated van.

It is important to use milk cans or bulk tankers that can be thoroughly cleaned and sanitized when transporting milk. Small-scale milk producers often transport their milk in cans, either by themselves or with the help of milk collectors, who act as unofficial dealers and intermediaries. In the process of milk collection, milk cans are collected from various producers and transported to small-scale processing facilities, family stores, stalls, canteens, and local/urban marketplaces using different modes of transportation such as bicycles, animals, cars, or by foot. The transportation process plays a crucial role in ensuring the delivery of high-quality milk, as milk in cans is not refrigerated during transportation. One advantage of utilizing milk cans is that they help preserve the quality of milk by preventing the mixing of milk from various producers (Djekic et al., 2018).

Large-scale commercial dairy farms commonly utilize bulk tankers for transporting their milk. Due to the insulation of milk tankers, the milk remains cold during transportation, reducing the risk of spoilage before reaching distant markets or processing facilities.

Transportation has a significant impact on the shelf life of dairy products due to their tendency to spoil more quickly compared to other foods (Figure 8.3).

Figure 8.3. *The logistics of bulk transportation of milk and dairy products.*

Source: Thrive (2023).

8.5.1. The Significance of Transporting Dairy Products

It is crucial to select the right vehicle equipped with temperature-controlling systems. Refrigerated vans offer a practical solution with their integrated refrigeration units and temperature monitoring systems. At Cool Running Rental, our fleet of refrigerated vans is equipped with top-of-the-line GAH Super Rapier SRF 21 Direct drive refrigeration units. These units are capable of reaching impressively low temperatures of –18°C (Cannas et al., 2020). Furthermore, the units are equipped with in-cab controls, allowing for easy temperature monitoring and adjustment right from the front seat.

Ensuring the proper maintenance and regular servicing of a van is crucial for the efficient delivery of dairy products. When a car breaks down, it becomes the least of concerns for a delivery driver who is out and about. Milk has a rather limited shelf life, so it is best to minimize its transportation whenever feasible. When a van experiences a breakdown, the potential outcome of it not being recovered can have a significant impact on whether or not it reaches its intended destination. When retailers, delivery drivers, and the farmers who grow the milk are affected, it leads to financial losses. Having an emergency plan in place is crucial for minimizing waste and preventing revenue loss in the event of breakdowns or traffic delays (Cannas et al., 2020).

Ensuring the safe transportation of dairy products necessitates strict compliance with sanitary regulations. Maintaining cleanliness and orderliness in your refrigerated van is of utmost importance. It is essential to ensure that both the produce handling spaces and vehicles are meticulously cleaned before each trip. Delivery drivers should undergo training to learn about proper hygiene protocols, including effective hand washing techniques and the correct way to wear protective gear. It is important for delivery drivers to carefully plan their routes in advance for every trip, considering any potential detours and traffic congestion. Ensuring prompt delivery necessitates the optimization of the delivery route to minimize travel time (Cannas et al., 2020). Effective communication regarding scheduled delivery times with customers and/or businesses would greatly enhance the safe transportation of dairy products. It enables individuals to align their receiving capacity.

8.6. REGULATIONS GOVERNING PACKAGING IN THE DAIRY INDUSTRY

Various environmental factors such as temperature, light, oxygen, microorganisms, and other variables have the ability to alter the composition of vitamins and minerals present in milk. Consequently, the packaging of dairy products must serve as a protective barrier to ensure the preservation of the food's nutritional content and its durability on the shelf (Biegańska, 2022). Food regulatory authorities have stringent regulations governing objects that come into contact with food.

Hygiene is of utmost importance in the domain of dairy packaging. Minimal alterations in the composition can lead to reactions that provide health risks. Aseptic packaging technologies tackle various concerns related to the dairy product, its packaging, and the filling process. Furthermore, the materials used adhere to quality regulations in the dairy product packaging industry. The materials must be properly engineered for the purpose of preserving and storing dairy products, and they must not contain any dangerous components (Biegańska, 2022). Specifically, the application of composite foils during pasteurization requires the employment of a specialized heat-resistant adhesive. Sterilization further augments the material's need for heat resistance.

The packaging for cooking cream (whipping cream), coffee cream, and milk needs to be durable, resistant to shocks, and completely sealed to prevent any leakage. To maintain stability, it is advisable to prevent the absorption of moisture from the liquid that is being wrapped. The shelf life of liquid dairy products affects the choice of packaging in terms of its barrier efficacy.

The label affixed to the packaging must have the following information: the identification of the product, the name and address of the producer, packer, or distributor, and the net quantity of the contents.

Furthermore, various agencies are engaged in the enforcement of the regulations. Additionally, laboratory testing may be required for packaging, similar to the requirement for things to meet criteria and be deemed safe (Biegańska, 2022). This is essential for organizations who desire to operate in a sustainable manner. Testing businesses can provide you with expert advice on the applicability of laws in a specific state. Alternatively, you could consult with legal professionals. It is advisable to get professional assistance and ensure compliance with all legislation. Regard this page as a manual, but bear in mind that it cannot substitute for professional advice.

8.6.1. The Fair Packaging and Labeling Act (FPLA)

This is a legislation that regulates the packaging and labeling of consumer products. The purpose of the FPLA is to simplify price comparison shopping for consumers and prevent misleading labeling and packaging of products such as pharmaceuticals, food, and cosmetics. Some categories and non-consumable items, such as tools, are not included in its coverage (Biegańska, 2022).

The legislation mandates that the packaging label must have a statement of identification, a declaration of responsibility (containing the name and address of the maker, packer, or distributor), and a declaration of net quantity, servings, or uses.

- *Give precedence to materials that are long-lasting:* The fundamental purpose of packaging is to ensure the protection of the product. Furthermore, employing resilient materials will assist in achieving this goal. Two exemplary instances of this are kraft paper and corrugated cardboard.
- *Acknowledge the expense associated with the materials*: It is important to consider the environmental impact of your packaging materials in relation to their cost. Although traditional materials like plastic may have greater availability and lower cost, they are also subject to taxes, such as the recently introduced Plastic Packaging Tax (Caban & Kravchenko, 2018). Moreover, the correct disposal of plastic might pose greater difficulties, as a significant portion of it ultimately finds its way into landfills or oceans.
- Aside from product quality, it is imperative for packaging supplies to possess excellent client service. Instead of simply shopping for products, seek out a strategic partner supplier that can assist you in maximizing the company's budget. Superior customer service can effectively address your packing issues. Products must adhere to environmental standards and comply with all relevant logistics and shipping rules and regulations. Furthermore, it is crucial to ensure that the things you purchase do not violate any current copyrights

and patents (Caban & Kravchenko, 2018).

- Ultra-high temperature pasteurization is employed during the production of aseptic milk to eradicate any possibly perilous microorganisms. Subsequently, the milk is promptly enclosed in a germ-free environment and stored in specialized packaging that prevents any contact with air. The method ensures the stability of the milk's shelf life.

Brand personality, commonly referred to as "Brand integrity," is formed by customers' evaluations of a firm or brand's visual presentation, reputation, and offerings. Effective packaging design should accurately convey the brand's values and evoke the desired emotions in buyers towards the products, hence effectively communicating the brand's offering (Caban & Kravchenko, 2018).

8.7. SUSTAINABLE PACKAGING SOLUTIONS FOR DAIRY PRODUCTS

The need for sustainability is more important than ever, and companies everywhere are actively searching for environmentally friendly alternatives to reduce their influence on the environment. The dairy industry, a major participant in the food sector, has also learned the importance of sustainable practices, particularly with regard to packaging (Rejeesh & Anto, 2023).

Conventional dairy product packaging, which includes single-use plastic containers and non-recyclable materials, has contributed significantly to environmental pollution and waste. These packages often end up in landfills or the ocean, where they take hundreds of years to break down and harm species (Rejeesh & Anto, 2023). Sustainable substitutes are obviously needed, and this presents an opportunity for the dairy sector to lead the way in enacting eco-friendly regulations.

8.7.1. Packaging Made with Biodegradable Materials

Biodegradable packaging composed of renewable resources such as cornstarch, sugarcane, or plant-based polymers is one of the most promising solutions. These materials have a very little ecological imprint because they naturally break down in the environment. Biodegradable dairy packaging not only minimizes waste but also conserves resources and emits fewer greenhouse gases during production (Rejeesh & Anto, 2023).

8.7.2. Compostable Packaging

This is another environmentally friendly option for the dairy sector. Compostable packaging, in contrast to biodegradable materials, breaks down into nutrient-rich compost, offering a sustainable means of replenishing the soil with essential nutrients. In a few of weeks, industrial composting facilities can turn this packaging into compost, helping the environment and completing the waste disposal process.

8.7.3. Repurposed Containers

Recycled packaging is crucial to the circular economy, which is becoming more and more popular. By utilizing recycled materials, dairy product packaging can drastically reduce the industry's dependency on energy and virgin resources. Paperboard and recycled plastics can be processed to preserve their quality and functionality, making them ideal for environmentally responsible packaging solutions (Rejeesh & Anto, 2023).

8.7.4. Containers That Are Refillable and Reusable

One unique opportunity to drastically minimize packaging waste is to use refillable and reusable containers. As part of a deposit-refund scheme, customers can return empty containers to dairy firms for cleaning and reuse (Rejeesh & Anto, 2023). This tactic encourages consumers to take responsibility for their actions and actively participate in sustainable practices while also reducing the amount of packaging waste produced.

8.7.5. Innovative Materials

Scientists and inventors are always looking into new environmentally friendly materials. There are a lot of interesting and diverse choices available for packaging that is friendly to the environment. For example, dairy products can be packaged using seaweed or mushroom mycelium. A lot of these innovative materials are safe and ecologically friendly because they are biodegradable, compostable, and sometimes even edible (Rajesh & Subhashini, 2021).

In order to guarantee a future for the world that is more ecologically friendly, industries must now see sustainability as a must. The dairy industry can make a big contribution to this effort by packaging its goods sustainably. Biodegradable, compostable, recyclable, and reusable materials as well as newly created materials can be used in place of conventional packaging techniques.

Dairy businesses who make investments in eco-friendly packaging will benefit from a competitive advantage in the market as consumers prioritize making eco-conscious choices in addition to contributing to environmental preservation (Rajesh & Subhashini, 2021). Collectively, the dairy business can influence other industries, create a more sustainable future, and make a difference by implementing these practices.

Finding methods to lessen the effects on the ecosystem is crucial, especially because environmental issues including ozone depletion, global warming, and unchecked soil and water pollution have become more prevalent in recent years. The fact that human-made products such as dairy product packaging have a significant impact on the environment over their entire life cycle should not be overlooked. Dairy packaging has an environmental impact throughout its whole life cycle, including the procurement and transfer of resources needed for product manufacture, distribution to supermarkets, final consumption, and waste management. This impact does not only occur during the product's manufacture, use, or disposal (Rajesh & Subhashini, 2021). For this reason, it's critical to incorporate environmental features and factors into the production of packaging for dairy products.

CHAPTER 9

DAIRY BY-PRODUCTS

Contents

The dairy industry generates a diverse array of by-products, including buttermilk, whey, ghee, and skim milk. Additionally, it processes raw milk into various products such as butter, cheese, cream, yogurt, condensed milk, dried milk, and ice cream. These dairy by-products are utilized in various non-food and food businesses due to their high nutritional value (Pires et al., 2021).

Buttermilk, a secondary product of butter production, is commonly used either in its dehydrated state or as a liquid (fermented to produce chaas). Whey, a nutrient-rich by-product of cheese and paneer production, has been utilized in the creation of sports drinks and other beverages (Pires et al., 2021). In addition, whey is employed in the production of certain cheeses, such as ricotta.

Skim milk, a byproduct of cream production, has been utilized to create cheeses such as cottage and quark cheese, as well as flavored milks. Leftover ghee from the manufacturing process has been utilized to produce chocolates, biscuits, and candies. Casein and its derivatives are mostly used in the baking and confectionery industries. In addition to their application in the food industry, caseins and whey proteins (WP) are utilized in the production of packaging films (Pires et al., 2021).

9.1. WHEY PROTEIN

Figure 9.1. *Whey protein powders.*

Source: Kunik & Fischer (2024).

Whey protein (Figure 9.1) is the term used to refer to the protein that is present in the liquid part of milk, which separates from the solid curds during the process of manufacturing cheese. It is commonly used as a dietary supplement to enhance protein intake. Whey protein can augment the nutritious value of the diet and influence the immune system.

Whey protein is a composite of proteins that have been extracted from whey, the liquid residue generated during the cheese-making process. The proteins mentioned are immunoglobulins, serum albumin, β-lactoglobulin, and α-lactalbumin (Kilara & Vaghela, 2018). Despite not being a protein, glycomacropeptide is nonetheless the third largest component. Whey protein is commonly marketed as a dietary supplement and has been associated with many advantageous effects on health. Whey protein is commonly utilized for the purpose of muscle development and improving athletic capabilities. Furthermore, whey protein is employed for the treatment of various additional conditions, such as diabetes, weight reduction, and asthma, however most of them lack robust scientific evidence.

Certain individuals integrate resistance training with whey protein supplements to enhance the synthesis of muscle proteins in their muscles and promote the growth of lean muscle mass. Two further potential benefits are lowering cholesterol levels and facilitating weight loss (Kilara & Vaghela, 2018). Casein and whey are the two primary proteins present in milk. Whey protein can be derived from the production of cheese or extracted from the casein present in milk. Whey protein is considered a complete protein because it contains all nine essential amino acids. It contains a minimal amount of lactose. The consumption of whey protein has numerous benefits, and researchers continuously uncover novel possible therapeutic applications for it.

Whey protein offers advantages such as facilitating weight loss and promoting muscular development. A study published in Nutrition & Metabolism, which included 158 individuals, discovered that the group drinking whey saw a notable reduction in body fat and demonstrated better retention of lean muscle compared to those who consumed the control beverage (Kilara & Vaghela, 2018). A study conducted over a period of 12 weeks examined the effects of whey supplements on 70 overweight individuals, both men and women. The study analyzed many indicators, such as cholesterol and insulin levels. The study revealed a notable reduction in both total cholesterol and insulin levels at week 12 in the whey group, as compared to the control group (Kilara & Vaghela, 2018).

9.1.1. Potential Risks

Excessive amounts of whey protein can induce cramping and abdominal discomfort; certain persons with milk allergies may also have a specific allergic reaction to whey.

While moderate amounts of whey protein typically do not result in any adverse effects, excessive quantities might lead to diminished appetite, migraines, exhaustion, queasiness, and the development of acne.

Prolonged and excessive use of whey protein might lead to the development of acne. Whey protein is a rare dietary substance that does not have a natural equivalent (Solak & Akin, 2012). There are concerns among some individuals over the potential hazards of consuming nutritionally refined foods like these. This is because, although these foods have a high nutrient content, the proportion of protein is significantly higher compared to other nutrients.

Whey protein concentrate (WPC), whey protein isolate (WPI), and whey protein hydrolysate (WPH) are the primary variants of whey protein. WPC has a low fat and carbohydrate content. The protein content in WPC fluctuates according on its concentration. According to Solak and Akin (2012), lower quality concentrates often have 30% protein, while higher quality concentrates can have as much as 90% protein. Whey protein isolate (WPI) undergoes additional processing to eliminate any traces of fat and lactose. WPI generally consists of a minimum of 90% protein. Whey protein hydrolysate (WPH) is considered the pre-digested version of whey protein as it has already undergone partial hydrolysis, a necessary process for protein absorption in the body.

Whey is the residual substance that remains after milk undergoes coagulation, which can be achieved by introducing acid or rennet. The solution consists of 5% lactose dissolved in water, together with a little amount of lipids and water-soluble proteins from milk. The available processing methods include either simple drying or the extraction of lactose, lipids, and other non-protein substances to enhance the proportion of protein content (Solak & Akin, 2012). The predominant technique for extracting protein from whey is membrane filtration, which can be accomplished by two methods: firstly, by subjecting the whey to "microfiltration," which effectively obstructs bacteria, casein micelles, and fat; secondly, by subjecting it to "ultrafiltration" (UF), which selectively hinders proteins. In addition to being an excellent protein source, whey also contains other beneficial components. Alpha-lactalbumin, beta-lactoglobulin, lactoferrin, and immunoglobulins are instances of this. Whey protein provides other supplementary health benefits, apart from promoting muscular growth, strength, and leanness. This involves reducing hypertension, hyperglycemia, and symptoms of stress and sadness. Additionally, it increases the lifespan of mice, reduces symptoms of hepatitis, enhances immune function in HIV patients, improves bone mineral density, and prevents cancer. The health advantages of whey protein seem to be influenced by its high cysteine concentration. Cysteine achieves this by elevating the amounts of glutathione, which serves as the principal antioxidant within the cells of your body (Solak & Akin, 2012).

Bodybuilders and athletes who aim to simultaneously build muscle mass and strength while reducing body fat will benefit greatly from incorporating whey protein into their daily diet. Whey protein serves as a very nutritious and effective method to enhance their overall protein consumption. According to Solak and Akin (2012), it is a protein source of superior quality that can be easily absorbed and utilized by the body. Protein is the primary and crucial ingredient for muscle growth and fat loss, and whey protein might even surpass other high-quality protein sources in terms of effectiveness.

9.2. LACTOSE

Lactose, also known as milk sugar, is a white, water-soluble solid that makes up approximately 2–8% of milk by mass. It is a disaccharide composed of glucose and galactose. Lactose is produced in large quantities as a by-product of the dairy industry, particularly from whey, which is the liquid left over after milk is curdled and strained. Lactose is purified from whey permeate or whey filtrate, and it can also be isolated by diluting whey with ethanol. Lactose is used in various applications, including pharmaceuticals, food processing, and fermentation. In the pharmaceutical industry, lactose is commonly used as an excipient, which is an inactive substance that helps facilitate the distribution of active ingredients in medications. It is found in about 60–70% of pharmacological dosage forms, including tablets, capsules, syrups, creams, and pastes. Lactose serves as a filler, providing volume and improving the flow of drugs. It also acts as a binder, helping the other ingredients in a tablet to blend and adhere to one another.

In food processing, lactose is used for its ability to convey flavors and colors. It is added to baked goods, seasonings, ice cream, dry soups, condensed milk, chocolate and chocolates, meat products, canned fruit and vegetables, coffee creamers, and skim milk. Lactose can control sweetness and lower costs when added to food.

Lactose is also involved in the process of fermentation, which is used to produce foods like cheese, yogurt, kefir, and acidified milk. Fermenting lactose produces lactic acid, which is used in the food, cosmetic, and pharmaceutical industries. The fermentation process involves adding lactic acid bacteria and sometimes yeast to milk or a milk product.

In addition to its applications, lactose has important roles in the human body. It is the only source of galactose, while glucose can be found in various foods. Galactose is a component of several macromolecules that are important constituents of nerve cell membranes and blood cell molecules that determine ABO blood types. Human milk contains 7.2% lactose, which can supply up to 50% of an infant's energy needs.

However, many people have difficulty digesting lactose due to a decrease in the production of the enzyme lactase, which breaks down lactose into glucose and galactose. This condition is known as lactose intolerance. Undigested lactose can be utilized as a nutrient by the intestinal microbiota and may have a lifelong effect of preventing the aging-related decline of certain immune functions. Lactose may also be involved in the absorption of calcium and other minerals, particularly during infancy.

In summary, lactose is a versatile substance with various applications in pharmaceuticals, food processing, and fermentation. It is produced in large quantities as a by-product of the dairy industry and is purified from whey. Lactose is used as an excipient in medications, providing volume and improving the flow of drugs. It is added to food to control sweetness and lower costs. Lactose is also involved in the fermentation process, which is used to produce foods like cheese and yogurt. Additionally, lactose has important roles in the human body, including as a source of galactose and as a nutrient for the intestinal microbiota. However, lactose intolerance can result in poor lactose digestion due to a decrease in the production of the enzyme lactase.

9.3. CASEIN

Figure 9.2. *Casein is a complete animal protein.*

Source: TruVitals Team (2023).

The protein in milk called casein is responsible for the milk's white hue. Casein protein makes up around 80% of cow's milk. Casein protein (Figure 9.2) can also be present in infant formulae, yogurt, cheese, and a range of dietary supplements in addition to milk. Casein peptides and casein protein are not the same thing. Casein protein is broken down into smaller bits to create casein peptides. Oral casein protein supplementation is used to treat a variety of illnesses, including diabetes, alcohol-induced liver damage, sports performance, and many more, although these claims are not well-supported by research (Horne, 2020).

The body gets all the amino acids needed to aid in muscle growth from casein protein. Since casein protein digests more slowly than other proteins, it may have a greater effect on satiety and appetite suppression. The body gets all the amino acids needed to aid in muscle growth from casein protein. Since casein protein digests more slowly than other proteins, it may have a greater effect on satiety and appetite suppression.

A large amount of proline amino acids found in casein prevent proteins from forming common secondary structural motifs. Moreover, disulfide bridges do not exist. It has comparatively little tertiary structure as a result. It is not very soluble in water due to its considerable hydrophobicity. Casein micelles are a suspension of particles found in milk that resemble surfactant-type micelles just somewhat in that they are spherical and have hydrophilic portions that are at the surface (Horne, 2020). Nonetheless, the core of a casein micelle is extremely hydrated, in stark contrast to surfactant micelles. Hydrophobic interactions and calcium ions hold the caseins in the micelles together.

A variety of molecular hypotheses could explain casein's unique structure in the micelles. One of them suggests that many submicelles comprise the micellar nucleus, with κ-casein microvillosities making up the periphery. According to a different concept, casein-interlinked fibrils create the nucleus. Lastly, a twofold link between the caseins is suggested by the most recent model in order for gelling to occur. Micelles are viewed by all three models as colloidal particles made of soluble κ-casein molecules encased in casein aggregates (Horne, 2020). Casein has a negative charge in milk because the pH of milk is 6.6. The refined protein has no solubility in water. It is easily dispersible in diluted alkalis and salt solutions such aqueous sodium oxalate and sodium acetate, despite the fact that it is likewise insoluble in neutral salt solutions.

9.3.1. Applications

- **Apply Paint:** Artists employ casein paint, a water-soluble, fast-drying material. Since tempera paint was first used in ancient Egypt, casein paint has been a popular choice for commercial illustrators. However, the introduction of acrylic paint in the late 1960s caused casein's popularity to decline (Horne, 2020). Scenery painters continue to utilize it extensively, while acrylic is also becoming more popular in that industry. Making casein adhesive

The ingredients of casein-based glues include casein, water, and alkalis (often a combination of sodium hydroxide and hydrated lime). After skimming the milk to remove fat, it is soured to cause the casein to precipitate as milk curd. After washing to get rid of the whey, the curd is pressed to remove any remaining water it might even be powdered. To manufacture glue, casein is combined with an alkali (often calcium hydroxide and sodium hydroxide) (Horne, 2020). Diverse alkali mixtures result in diverse qualities for glues. Moreover, preservatives might be applied.

- **Food:** A range of caseinates can be found in a number of foods, creams, and toppings. Although calcium caseinate has a higher calcium content and a lower salt content, firms may choose to utilize sodium caseinate as a food additive to stabilize processed foods. Proteins and fat from milk typically the milk of cows, buffalo, goats, or sheep compose cheese. It is created by coagulation, which starts the fractionation and selective concentration processes by causing the casein micelle to become unstable. Usually, rennet which contains the proteolytic enzyme rennin is added to milk to acidify it and cause it to coagulate. Rennet was formerly made from the stomachs of calves, but it is now more frequently made from genetically altered microbes (Horne, 2020). After that, the solids are divided and formed into their final shape.
- **Supplementation of Protein:** The capacity of the casein molecule to coagulate or gel in the stomach, which increases its efficiency in supplying nutrients, is one of its appealing qualities. Amino acids can be slowly released into the bloodstream by the clot, sometimes for up to many hours. Casein is frequently sold as hydrolyzed casein, which has been broken down by a protease like trypsin. Because hydrolyzed forms are known to taste bitter, babies and laboratory animals frequently reject these supplements in favor of intact casein.

A small percentage of people have a casein allergy. When the body is unable to digest the casein proteins, it results in casein intolerance, sometimes referred to as "milk protein intolerance." In early children, the prevalence of casein allergy or intolerance varies from 0.25 to 4.9%. Adult and older child numbers are unknown. Many people on the autistic spectrum develop an adult-onset allergy or intolerance to casein protein (Horne 2020). Clinicians and dietitians can utilize this to identify autism in people who might not exhibit typical autistic features. These people typically follow a casein-free, gluten-free (CFGF) diet after learning they have an allergy or sensitivity.

It has been demonstrated that casein that has been heat-treated is more allergic and more difficult for babies to digest. Although breast milk has not usually been demonstrated to induce an allergic reaction, it should always be given to a baby with caution if the nursing parent has taken something that contains casein. It has been demonstrated that breastfeeding infants who are allergic to or intolerant of dairy protein does better when they follow a diet free of casein. When available, human breast milk should be tried first because it has been shown to be the best nourishment for infants.

9.4. BUTTERFAT

Figure 9.3. *Milk butter.*

Source: Shani (2022).

The lipid component of milk is referred to as butterfat or milkfat. The label often provides information on the quantity of butterfat present in milk and cream. Butterfat is mostly composed of triglycerides. Each triglyceride has three fatty acids (Figure 9.3).

Trans fat constitutes around 3% of the total butterfat content, which is little less than 0.5 grams per US tablespoon. Ruminant meat and milk contain naturally-occurring trans fats. The predominant kind of trans fat found in milk is vaccenic fatty acid (Li et al., 2010). Some commercially produced foods, such as shortenings prepared from hydrogenated vegetable oils, may also contain trans fats. Nutritional experts advise minimizing trans-fat intake to negligible levels because all trans fats provide an identical risk to health, as supported by established scientific studies.

Butterfat is the primary constituent of butter and is a naturally-occurring fatty substance found in cow's milk. Anhydrous butterfat has a slower rate of rancidity compared to butter and can be stored at room temperature for several months. When butter is melted, the pure butterfat floats to the surface and may be separated, while the remaining curd and water promote the growth of microorganisms that cause the butter to spoil. Butterfat is a prevalent component in gourmet foods and is utilized in culinary applications (Li et al., 2010).

Butterfat is mostly composed of a mixture of triglycerides, particularly those derived from fatty acids such as stearic, palmitic, oleic, and myristic acids, in the context of chemistry. The fatty acid content of butterfat is influenced by the food of the animal that generates it.

The variation in hue observed in cheese is also ascribed to the presence of butterfat. The concentration of beta carotene in milk fluctuates based on the season and the composition of nutrients present in the grass. The hue of milk fluctuates in accordance with the season; during the winter, it appears nearly white, while in the spring and summer, it takes on a luscious, creamy golden tone. Butterfat is a vital element in determining the quality of cheese. Increased butterfat level is linked to a more intense flavor and a smoother texture. Conjugated linoleic acid (CLA) is a beneficial nutrient that aids in reducing abdominal fat and enhancing muscle growth. Vitamin K2 is another important nutrient that supports the health of the thyroid and heart. Additionally, vitamin A is also beneficial for overall health (Li et al., 2010).

9.4.1. Health Benefits

Butter is high in calories and fat, but it also contains numerous essential components. For example, it is an excellent source of vitamin A, a vitamin that dissolves in fat and is necessary for robust immune function, vibrant skin, and sharp eyesight. Additionally, it contains vitamin E, which supports cardiovascular health and acts as an antioxidant, protecting cells from damage caused by free radicals (Li et al., 2010). In addition, butter contains minimal amounts of niacin, calcium, phosphorus, and riboflavin.

Butter is rich in conjugated linoleic acid (CLA), a type of fat that is found in meat and dairy products. CLA has been linked to significant health benefits.

Butyrate, a type of short-chain fatty acid associated with various benefits, is found in high quantities in butter (Shin et al., 2013). Furthermore, butyrate, which is synthesized by beneficial bacteria in the gastrointestinal tract, serves as a source of energy for the cells lining the intestines. By reducing inflammation in the intestines and promoting the absorption of fluids and electrolytes to preserve regularity and balance, it can enhance digestive health. Moreover, it may alleviate the symptoms associated with irritable bowel syndrome (IBS), such as diarrhea, constipation, bloating, and abdominal pain. Research suggests that butyrate may possess anti-inflammatory properties when used in the treatment of Crohn's disease. Butter is rich in saturated fat, which is a type of fat found in foods such as meat and dairy products. Butter is primarily made up of approximately 63% saturated fat, 26% monounsaturated fat, and 4% polyunsaturated fat. Historically, saturated fat was commonly believed to be a detrimental form of fat that obstructed arteries and had negative implications for cardiovascular well-being (Shin et al., 2013). Current research indicates that there is no evidence linking the consumption of saturated fat to an elevated risk of developing heart disease or mortality from it.

Butyric acid is a low molecular weight fatty acid that is present in butter. Butter is the primary dietary source of butyrate, with a concentration of approximately 3% to 4% butyric acid. Research studies have demonstrated the efficacy of butyric acid in promoting weight loss. It operates by promoting the growth of beneficial bacteria in the stomach.

Butterfat consists of saturated, monounsaturated, and polyunsaturated fats. According to the American Heart Association, it is recommended to consume 5% to 6% of total calories from saturated fat (Shin et al., 2013). Thus, a diet consisting of 2,000 calories may include the consumption of 13 grams of saturated fat per day. Although it is not advisable to regularly consume excessive amounts of saturated fats, you can indulge in foods rich in butterfat, like butter and ghee, in moderation, as long as you maintain a balanced diet. In addition to their high fat content, foods rich in butterfat, such as ghee and butter, also include a significant amount of essential vitamins and antioxidants that are important for a balanced diet.

The primary factors that influence the quantity of fat in cow's milk are the breed and lineage of the dairy cow, food patterns, age, and overall animal management and milking protocols. Hence, the level of care provided to the cow greatly influences the fat content of its milk, regardless of its age and pedigree. Organic Valley cows are raised in a humane manner, without the administration of antibiotics, while grazing on pastures. Due to the optimal conditions, the dairy cows naturally yield milk with a higher fat content. Superior dairy products such as Organic Valley Butter, Organic Valley Ghee, and Organic Valley Whole Milk can be produced from this milk that is rich in butterfat (Shin et al., 2013).

The overall flavor and consistency of a food are greatly impacted by the amount of butterfat it contains. Whole milk offers a more indulgent flavor and has a denser consistency in comparison to skim milk. Cream products with a greater fat level, like heavy cream, may taste greasy and excessively heavy when drunk alone due to their extraordinarily high butterfat content (Shin et al., 2013). A minimum of 40% of the butterfat in Organic Valley Heavy Cream can be easily whipped into smooth peaks. The rich flavor of this ingredient is also very suitable for inclusion in baked goods or as a creamy addition.

9.5. DAIRY BY-PRODUCT UTILIZATION

The dairy business produces a variety of by-products such as buttermilk, whey, ghee, and skim milk in addition to processing raw milk into a wide range of goods such as butter, cheese, cream, yogurt, ghee, condensed milk, dried milk, ice cream, etc. Due to their great nutritional value, these dairy by-products are used in both non-food and several food industries. As a by-product of producing butter, buttermilk can be utilized as an ingredient in both liquid and dried form. Sports drinks and other beverages have been made with whey, a high-nutrient by-product of the production of cheese (Shah & Patel, 2019).

Additionally, whey is utilized in the making of some cheeses, such as ricotta. Cheeses like cottage and quark cheese, as well as flavored milks, have been made from skim milk, a byproduct of making cream. Chocolates, cookies, and candies have all been made from the leftover ghee from the manufacturing process (Shah & Patel, 2019). The primary use for casein and its derivatives are in baking and confections. Apart from their use in food, caseins and whey proteins (WP) are also used in packaging films.

9.5.1. Whey Utilization

Despite the fact that these whey proteins are added to sports drinks and other agro-food products, 40% of whey is remained raw, which is an intriguing resource given its superior oxygen barrier qualities. Excellent oxygen, fragrance, and oil barrier qualities are found in whey protein films. When applied as coatings on food products, films dividing layers of different meals, or films molded into food ingredient pouches, their superior mechanical qualities offer durability (Shah & Patel, 2019). These films bring value for final commercial applications since they do not materially impair the desired main barrier and mechanical qualities as packaging films.

Whey-based compositions are processed via compression molding and extrusion for use in edible coatings and packaging applications. To get around the inherent brittleness, plasticizing chemicals must be used. Isolate films of whey protein are completely translucent. There are now whey coatings that include both an active and a barrier layer. To increase the shelf life of the packed food, the active layer comprises antioxidants or antimicrobials, and the barrier layer contains isolates of whey protein enhanced with plasticizers. Whey-based films have the potential to enhance the coated goods' sensory qualities while offering consumers some health advantages (Shah & Patel, 2019).

9.5.2. Ghee Residue Utilization

Ghee is clarified milk fat with unparalleled organoleptic qualities that make it a vital element in a wide range of food applications. It is a significant component of Indian cuisine cooked in various ways. In the food industry, ghee residue is used to make baked goods, desserts, and flavor enhancers. The nation produces a considerable amount of ghee residue (GR), which is a highly nutritious source of proteins and nitrogenous substances. Burfi is made by combining leftover ghee with sugar, cocoa, and skim milk powder (SMP). It is useful for making toffee, coconut burfi, and candies after combining with other ingredients (Shah & Patel, 2019). Use the nutrient-rich byproduct as a food additive in a range of dishes, soups, and spreads.

9.5.3. Buttermilk Utilization

Buttermilk fortification is used to boost production as long as the quality of the cheese is maintained. Buttermilk is an interesting dairy ingredient for use as a functional ingredient in a variety of food products like chocolate, cheese seasonings, margarine, bread, ice cream mixes, or yogurt. Its high phospholipid content also plays a significant role as emulsifiers in food systems. To enhance the organoleptic, rheological, and functional qualities of processed cheese spread, buttermilk concentrate (BMC) rich in phospholipids has been added (Sar et al., 2022). Processed cheese spreads are made beneficial as functional foods since they include BMC. This approach may also result in cost-effective revenue by increasing product yield or employing dairy sector byproducts with low value, such buttermilk.

9.5.4. Skim Milk

A company based in Egypt attempted to produce Domiati cheese earlier, but the cheese produced was not as good as that made from fresh milk. The recombined cheese had a firm texture and developed weak flavor even after the normal maturation period. Later attempts were made to improve the quality of recombined Domiati cheese, and after a month of storage, they were able to achieve an acceptable taste and flavor by directly adding skim milk powder up to 35% total solids to buffalo or cow milk and coagulating the concentrated milk covered with brine (Sar et al., 2022).

9.6. ENVIRONMENTAL IMPACT OF DAIRY BY-PRODUCTS

The dairy business is accountable for various forms of environmental pollution, such as substantial emissions of greenhouse gases that worsen climate change, due to its industrial-scale operations that house thousands of cattle. Not only does the industry harm the environment, but it also causes a great deal of suffering to animals. Littler, family-run farms are being forced out of business by the few giant multinationals that have come to control the industry (Pardo et al., 2016). Smaller farms find it difficult to supply milk at the low prices that large farms, which are sometimes supported by subsidies and other financial incentives, set.

Because human activity produces more greenhouse gases (GHG) and the potential for climate change than the earth's natural carbon and nitrogen cycles can absorb, an analysis of food systems is necessary to determine where in the food chain GHG are produced and to help make decisions for reducing them. An estimated 21% of all GHG from human activity is related to the growing, processing, transportation, consumption, and disposal of food (Pardo et al., 2016).

The majority of greenhouse gas emissions occur from emissions prior to milk production up to the farm gate, according to reports from the food and agriculture organization (FAO) and other studies on milk. The rumen activity of cows is linked to a significant amount of greenhouse gas emissions from milk production. Reducing the greenhouse gas emissions from cows will be a major challenge in mitigating methane because of operational and economic issues. Even if the entire GHG from the milk life cycle is substantially reduced by dairy processing, process improvements can be made with the use of current technology (Pardo et al., 2016). Therefore, reducing the greenhouse gas impact of dairy processing is more an economic issue than a technological one.

Air contamination In the United States, the dairy industry contributes significantly to greenhouse gas emissions. Fifteen percent of the nation's estimated greenhouse gas emissions are attributed to dairy farms. In addition, dairy farms are accountable for a significant portion of the nation's ammonia emissions, ranging from 19 to 24% (Pardo et al., 2016). Pollution from dairy farms and other livestock production facilities can be fatal. Actually, the number of deaths linked to livestock air pollution now exceeds the number of deaths linked to coal-fired power stations. Emissions from dairy farms are responsible for about 2,000 fatalities (Capper et al., 2009).

Pollution of Water Waterways in the neighboring areas are contaminated by intensive dairy farming operations, rendering them unsafe and useless. Large vats are used to hold the manure from thousands of dairy cattle kept in factory farms until it is time to spread it on neighboring crop fields. However, the amount of manure is simply too great to apply to the field in a safe and useful manner, thus nitrogen and phosphorus frequently leak into neighboring streams. Over time, these vats may also break and tear, allowing their contents to leak out and seep into groundwater and reach both tiny and large bodies of water. During its production into an edible product, most milk undergoes at least one thermal processing step. As a result, both the energy input and the additional resources required to clean the heat exchangers that are prone to fouling and the remaining plant processing equipment are substantial (Capper et al., 2009). Enhancements in heat exchanger cleaning could result in a notable decrease in energy consumption and ecological footprint. A review of cleaning heat exchangers is presented, with the conclusion that cleaning linked to fouling increases process installation capital requirements, energy and water demands, labor costs, and lost processing time. Fouling during milk processing has very little effect on pumping needs; estimates of category contributions to the overall cleaning cost of water, cleaning agents, labor, and energy are 25.6, 23.2, 23.9, and 27.3%, respectively (Capper et al., 2009).

CHAPTER 10

FUTURE OF DAIRY INDUSTRY

Contents

Dairy products, including milk, have been a major source of vital nutrients for mankind. However, the ever-changing demand and consumption for them have always had an impact on dairy farming and milk production.

The dairy industry must contend with a number of environmental, political, and economic issues that further influence its growth in addition to the shifting requirements and habits of its customers (Fox, 2012). Every industry is undergoing a technological revolution, and the dairy industry is no exception. People are seeing how different cutting-edge technologies are being integrated to change conventional dairy processes. The following significant technical developments will influence dairy farming in the future:

- **Automated Milking Systems:** These devices simplify the milking procedure, cut down on labor needs, and enhance the quality of milk produced (Fox, 2012). Robotics is used in these systems to monitor, milk, and clean cows, guaranteeing uniformity and efficiency.
- **IoT and Data Analytics:** Dairy farmers can now check milk production, keep an eye on cow health, and improve breeding and feeding procedures through Internet of Things (IOT) and data analytics. For increased productivity and animal welfare, farmers may make more informed decisions with the use of sensors, wearable technology, and data-driven insights (Fox, 2012).
- **Precision Nutrition and Feeding:** To develop individualized feeding schedules, precision nutrition systems examine data on cow productivity, health, and dietary needs. Farmers may maximize environmental effect, cut expenses, and increase milk output by optimizing feed mix and rationing.

10.1. INNOVATIONS IN DAIRY PROCESSING

In the quickly changing field of agriculture, innovation is essential to both surviving and remaining relevant. This also applies to the dairy industry, a major player in the world food chain. In order to secure a sustainable and prosperous future, dairy enterprises are adopting a variety of cutting-edge methods and technology in response to the growing demand for dairy products and the growing worries about the environment.

In 2024, the dairy business is undergoing a period of rapid innovation as it uses cutting-edge technology and environmentally friendly procedures to respond to changing consumer expectations and trends (Gebeyehu, 2023). The processing, packaging, and delivery of dairy products have seen a substantial change this year, with a greater emphasis on effectiveness, quality, and environmental responsibility.

The use of automation and artificial intelligence is a major development in the dairy sector. Dairy processing is being revolutionized by these technologies, from pasteurization and milking to packing. Automated systems improve manufacturing processes, lower the possibility of human mistake, and increase operational efficiency. Artificial Intelligence is being applied to supply chain logistics optimization, maintenance demand prediction, and quality control.

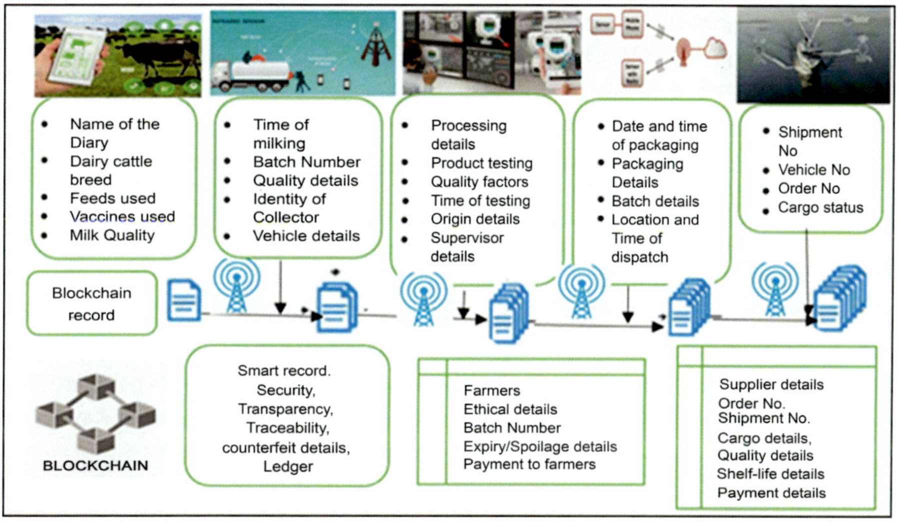

Figure 10.1. *Dairy supply chain integrated with emerging technologies and blockchain.*

Source: Author4o (2023).

Innovations in the industry are leading the way in sustainability. Eco-friendly methods, like employing renewable energy sources, conserving water, and lowering carbon footprints, are being used by dairy processors more and more (Gebeyehu, 2023). This change responds to customer demand for sustainable products as well as environmental concerns.

Leading provider of quality management software, RizePoint, is essential to this change (Gebeyehu, 2023). Dairy processors are able to control their environmental impact, adhere to strict food safety laws, and maintain high levels of quality thanks to their solutions. In particular, the software's supply chain supervision, audit management, and compliance tracking features help to guarantee sustainable and consistently high-quality products.

Additionally, the dairy business is seeing a shift in customer tastes toward plant-based and organic dairy alternatives. In response, dairy processors are expanding the range of products they offer and adding substitutes like oat, soy, and almond milk. These processors are assisted in preserving the quality of both conventional and non-traditional dairy products by RizePoint's adaptable quality management system(Gebeyehu, 2023).

Some other innovations include:

- **Mechanized Agriculture and Robotics:** Imagine a dairy farm where robots milk the cows, automated systems feed them, and sensors track their health. In contemporary dairy farming, this ideal is starting to come to pass. Streamlining operations and cutting labor expenses are autonomous feeding systems, robotic milking systems, and even drones for herd management. Now that basic activities are being handled by automation, farmers can concentrate on making strategic decisions.
- **Ecological Methods:** In agriculture, sustainability is a buzzword, and dairy farming is no different. Regenerative agriculture, organic farming, and water conservation are becoming more popular. These methods not only lessen the negative effects of dairy production on the environment, but they also appeal to environmentally concerned customers who wish to support companies that practice environmental responsibility (Nimbalkar et al., 2021).
- **Other Sources of Feed:** The dairy industry is investigating feed substitutes as the globe looks for sustainable sources of protein. Algae and insect-based protein are becoming more and more viable choices. These substitute feeds provide new opportunities for innovation while lowering the environmental impact of dairy production.
- **Blockchain Openness:** Transparency in supply chains is being revolutionized by blockchain technology. Blockchain technology in the dairy sector allows customers to follow the path of their dairy products from farm to table (Nimbalkar et al., 2021). Customers can make educated decisions thanks to this transparency, which also strengthens trust and guarantees product authenticity.
- **Plant-Based Alternatives to Dairy:** Plant-based dairy substitutes like cashew cheese, soy-based yogurt, and almond milk are on the increase as a result of the growing popularity of plant-based diets. In order to meet the demand for these alternatives, dairy businesses are responding by expanding their product offerings.
- **Research on Microbiomes:** There are □ignificant ramification□ for comprehending and improving the microbiota of dairy cattle. Farmers may increase milk production, lower disease outbreaks, and produce healthier herds by improving the digestive health of their cows (Nimbalkar, Verma, & Singh, 2021).

- **Creative Packaging:** The use of eco-friendly packaging is growing. In an effort to cut down on plastic waste, dairy companies are looking into reusable and biodegradable materials (Nimbalkar, Verma, & Singh, 2021). Environmentally conscious consumers are drawn to packaging that is both appealing and sustainable.
- **Nutrition-Boosting Dairy Products:** Foods with functional properties are becoming more and more popular. In response, dairy companies are creating products enhanced with probiotics and other health-promoting ingredients to meet consumer demand for wellness and health.
- **Models of Direct-to-Consumer:** Dairy sales and distribution are changing due to e-commerce and direct-to-consumer business methods. Online access to dairy goods is a recent development that is completely changing the way dairy businesses market to consumers.
- **Energy Effectiveness:** Energy-efficient solutions in dairy farms and processing facilities are lowering operating costs and their environmental impact (Nimbalkar et al., 2021). Dairy production is becoming more economically and sustainably feasible thanks to these developments.

The dairy sector depends on innovation to help it respond to shifting customer demands and environmental issues. These advances contribute to a more sustainable future in addition to increasing productivity and product quality. It's becoming increasingly evident that embracing innovation is essential for success as the dairy industry develops. With each new step it takes toward innovation, the sector is advancing toward a more promising and sustainable future.

10.2. FUNCTIONAL DAIRY PRODUCTS

Up until now, dairy products have led the way in the development of functional foods. Since fermented dairy products have long been associated with health benefits, it makes sense for the dairy industry to expand its product line to include other kinds of health-promoting goods. In recent years, the availability of functional dairy products has increased in the daily life that has become more and more popular (Mattila-Sandholm & Saarela, 2003). Consumers' concern for their own personal health has created markets for these products. In the near future, we will undoubtedly see more products designed for specific consumer groups. Milk is regarded as a practically complete food since it contains a significant amount of fat, protein, and essential minerals. Additionally, milk and its derivatives are essential components of a regular diet, particularly for vulnerable populations including infants, school-age children, and the elderly.

Aside from breast milk, milk is one of the most vital dietary sources for nourishment for newborns and young children. Dairy product consumption has really been connected in recent times to health advantages that directly combat diseases and complications associated with being overweight or obese.

Milk is a composite physiological fluid that provides nutrients and bioactive ingredients to help babies adjust postnatally during digestive maturation. It facilitates the growth of lymphoid tissues and the symbiotic microflora that they sustain. The quantity, potency, and significance of milk's bioactive components may be taken into account beyond previous research. They consist of proteins, bioactive peptides, oligosaccharides, phosphorus, selenium, magnesium, zinc, riboflavin, calcium, some particular organic acids, and carbs. Though occasionally these are ingredients in fresh milk, they primarily surface during fermentation or digestive processes. Inhibition of intestinal cholesterol absorption is one of the potential ways by which fermented dairy products and probiotic bacteria reduce or eliminate cholesterol (Mattila-Sandholm & Saarela, 2003).

For instance, people who eat dairy products are more likely to be overweight, to have lower blood pressure, to be healthier, to have a lower risk of stroke, colon cancer, and to have osteoporosis. A variety of functional foods have been developed recently, and many of them are produced globally. These include probiotic, prebiotic, and synbiotic foods, as well as foods enhanced with antioxidants, phytosterols, and foods reduced in fat, salt, or sugar (Mattila-Sandholm & Saarela, 2003).

Probiotic functional foods are among those that have improved general health. The probiotic dairy foods and probiotic non-dairy foods can be separated out. Every year, the probiotic dairy food business grows. The health benefits of probiotic bacteria, which were first discovered in milk products, the bioactive ingredients in fermented dairy products, and the ability to reduce lactose intolerance have all contributed to a surge in demand for dairy probiotic products (Mattila-Sandholm & Saarela 2003). As a result, creating these goods is a major area of focus for food design research and a problem for the scientific and industrial communities.

10.2.1. Dairy Products with a Purpose

Dairy products are well-known for being organic, healthful, and full of the essential components of a balanced diet. Dairy products have been used to "treat" high blood pressure, among other chronic conditions, in addition to their nutritional benefits (Saarela, 2007).

While the addition of probiotic microorganisms and prebiotic fibers is one of the most widely used main principles in the formulation of functional dairy products, other popular approaches include the use of minerals, phytosterols, conjugated linoleic acid, omega-3 fatty acids, and bioactive peptides to address a wide range of health issues, from maintaining gastrointestinal health to controlling hypertension, lowering cholesterol, and improving immune system response, among many other health concerns.

- **Functional Yoghurt:** One type of functional food is yoghurt, which is made through lactic acid fermentation primarily by the Streptococcus and *Lactobacillus* species. It has a variety of health-promoting properties, such as enhancing the gut microbiota, reducing lactose absorption and intolerance by improving gastrointestinal functions, boosting the immune system, improving lipid digestibility, and having antimicrobial properties. The number of these health-promoting properties of yoghurt can be increased by adding probiotic microorganisms, prebiotic ingredients, bioactive peptides, and other ingredients. While milk is said to contain significant amounts of polyphenols, several factors, like pasteurization and bacterial actions to milk protein, reduce the amount of polyphenolic compounds and naturally occurring antioxidants.
- **Probiotic Foods and Supplements:** Probiotic bacteria are live strains that, when taken in moderation, improve the microbial balance in the gastrointestinal tract and have positive health effects. People are interested in consuming probiotic-containing products because they have several beneficial effects, including: (a) improving the nutritional value of food products; (b) controlling and reducing the serum cholesterol; (c) improving the immune system; (d) preventing gut infections and suppressing antibiotic-associated diarrhea; (e) reducing lactose intolerance symptoms; (f) reducing the risk of colon cancer; and (g) improving the digestion of gliadin against celiac in gluten-containing foods that are gluten-containing (Saarela, 2007).
- Dairy proteins are essential for controlling food intake, feeling full, and preventing metabolic side effects associated with obesity. Calcium, peptides, milk proteins, and lactic acid bacteria can all have an impact on blood pressure. Some of the components in milk fat have physiological relevance. Sphingolipids and their active metabolites have antimicrobial properties that they might impart directly or through digestion. During the Middle Ages, whey was investigated as a skin balm, aphrodisiac, and medicinal (Saarela, 2007). Whey proteins, including α-lactalbumin, lactoferrin, serum albumin, lactoperoxidase, and β-lactoglobulin, have significant biological and nutritional qualities, especially in relation to disease prevention.

10.3. PERSONALIZED NUTRITION IN DAIRY INDUSTRY

The wide term personalized nutrition, also known as individualized nutrition, refers to a strategy that aims to give customers access to meals that suit their unique needs and/or wants for a healthier diet and way of life. Based on the idea that no two people are alike that is, that each of us is unique in our biochemistry, metabolism, genetics, and microbiota a personalized nutritional plan can be made that takes these aspects into account based on our individual eating patterns, body measurements, weight, cholesterol, and lifestyle choices (Porzi et al., 2021). Other names for personalized nutrition include "nutritional genomics," "precision nutrition," and "individualized nutrition," though these terms sometimes have similar meanings in the literature.

Personalized nutrition creates nutrition recommendations that is specific to you by taking into account characteristics like your metabolism and demands, unlike general, one-size-fits-all suggestions for healthy eating. Everybody reacts to food differently, and your chance of developing chronic illnesses is correlated with your response to food.

Food producers are aggressively supporting consumers with their nutritional needs in response to the growing demand for fitness and well-being (Porzi et al., 2021). In order to adapt to this changing environment, the dairy industry focuses on customized formulations and functional dairy products as specialized nutrition solutions.

10.3.1. Current Advancements in Dietary Specialization

- **Wholesome Health:** The dairy business, like all others, has realized that consumers are becoming more and more interested in holistic well-being, and it is making adjustments to fulfill those needs. Fonterra, a prominent dairy company, has launched two cutting-edge brands, BioKodeLab and Nutiani, with a strategic aim to improve muscle health, mobility, immunity, and cognitive wellbeing (Porzi et al., 2021).

The company has created products like Focus Plus and Calm Plus with BioKodeLab that are especially designed to assist brain nutrition. Conversely, Nutiani has introduced a line of products aimed at promoting people's overall well-being, focusing on three essential aspects: mental, physical, and interior well-being.

- **Mobility and Physical Performance:** It hasn't been the case that dairy is the preferred option for athletes. But there was a distinct change. Dairy businesses began to market their products as nutrient-dense, naturally occurring options for boosting athletic performance, preserving muscle health, and promoting general mobility.

- **Support for Immunity and Gut Health:** By marketing its products as a handy way to maintain healthy immunity and enhance general well-being, the dairy industry also hopes to improve gut health and boost consumer immunity.

Dairy companies will be able to stay ahead of the curve by attracting and retaining the interest and allegiance of health-conscious consumers looking for customized nutritional options thanks to the emphasis on specialty nutrition. Consequently, the dairy sector will keep seeing incredible advancements in the field of targeted nutrition (Simopoulos & Milner, 2010).

Another important factor in drawing more health-conscious consumers to the dairy isle is the active nutrition section. Similar to biotics, protein is a highly desired element for many consumers, and active nutrition products are reflecting this, according to Medina.

According to the Growth Opportunities for Dairy research, consumers believe statements made on packs about probiotics (55%) and protein content (58%) to be appealing. Blending traditional protein sources like whey and casein with soy and pea proteins, along with probiotics, is opening up new dairy consumption options for consumers who prioritize active nutrition. These opportunities range from supporting exercise to providing afternoon energy boosts. In response to consumer desire for quick, healthier meals, dairy-based drinks, frozen desserts, snacks, and confections with a variety of functional ingredients including additional protein, prebiotic fiber, probiotics, and postbiotics are starting to appear on the market.

Dairy manufacturers can benefit from expanding their product line to include plant-based protein sources such as nuts and seeds. Medina noted that consumers who prioritize active nutrition find a range of protein sources appealing, with whey and plant-based sources being the most popular. As a result, there is a clear opportunity for dairy manufacturers to champion both traditional proteins and plant proteins through blends (Simopoulos & Milner 2010). Nutritious ingredients like nuts and seeds can further support these protein blends, helping achieve consumer-preferred taste and texture as well as ingredient and protein diversity within a variety of dairy offerings.

Additionally, it aids in achieving clean-label goals. Although there isn't a single definition for a clean label, research reveals that the majority of customers say that simple, identifiable components have an impact on their decisions to buy products. Brands may draw in conscientious consumers who identify familiar, healthy components with qualities that are beneficial for them, their communities, and the environment by utilizing mixes with these ingredients.

Personalized diets and supplementation are in high demand as people's awareness of holistic well-being grows. People are shifting toward individualized diet as a result of increased consumer health awareness and greater disposable money across regional boundaries. Individually tailored diets are in high demand due to shifting consumer preferences for wellness and health products. A more individualized, genetically based diet is becoming increasingly popular as people become more discerning about their nutrition. Supplements are becoming less common, and people are switching to individualized nutrition regimens and supplements based on their genetic makeup to treat various health issues (Simopoulos & Milner, 2010). Customers are moving away from the "one-□ize-fit□-all" approach and wanting more cu□tomization in their everyday experiences as a result of their changing preferences brought on by greater health awareness.

10.4. CLEAN LABEL TRENDS IN DAIRY PRODUCTS

The dairy sector is faced with the difficulty of satisfying clean label regulations while preserving product safety and quality in response to growing customer preferences for minimally processed goods. Consumers that prefer easily recognized components over complicated chemical names are drawn to clean labels, which are characterized by a transparent and simple ingredient list. According to □urvey□, there i□a general under□tanding of the i□□ue; 76% of con□umer□think item□with clean label□are healthier. According to a □urvey, 76% of respondents said they were more likely to purchase items with components they could trust (Roobab et al., 2023). This suggests that customers' preferences for ingredients they feel confidence about influence their purchasing decisions.

Figure 10.2. *Natural and clean label trends.*

Source: foodnavigator-usa.com (2019).

Consumer decisions are influenced by worries about strange substances and titles that seem chemical, with 36% of consumers saying they are less likely to buy products that include strange ingredients.

Even while consumers want clean labels, the dairy business finds it difficult to switch off conventional synthetic preservatives. Investigating natural substitutes is essential to preserving food safety and prolonging shelf life without sacrificing consumer expectations. For instance, citric acid can be obtained organically from citrus fruits or fermentation by widely accepted safe organisms, offering a clean label option even though it is manufactured synthetically.

The food business and consumers have different ideas on what constitutes a clean label, and this interpretation is ever-changing. Foods with a clean label are usually identified by their ingredients, which emphasize natural flavors, colors, and preservatives while omitting artificial additives. Though there are minor variations in viewpoints, it is generally agreed upon that these foods consist of readily recognizable, natural constituents that have undergone minimum chemical processing (Roobab et al., 2023).

Foods labeled natural or organic, as well as those that support particular value claims like no bioengineered ingredients, no added hormones or antibiotics, sustainable practices, plant-based, fair trade, or eco-friendly packaging, are associated with the term "clean label," according to a number of sources. The concept of clean label is changing, emphasizing its link to products made with ecological and ethical standards, which is influencing customer expectations even more.

Consumer demand for transparency and health-conscious options is driving the shift toward clean label dairy. Recent market research firms' surveys show that over 70% of consumers actively seek out products with identifiable ingredients, and that 65% are willing to pay a premium for clean label dairy options. These figures demonstrate the significant demand for and willingness to pay extra for dairy products that are healthier and more transparent.

10.4.1. Challenges in Clean Label Dairy Formulation

Developing clean label dairy products comes with a unique set of difficulties. Keeping freshness, texture, and flavor without using conventional additives calls for a careful balancing act between science and creativity. One major obstacle is extending shelf life without sacrificing the product's clean label status. To get around these issues, dairy producers are experimenting with novel ingredient combinations. Natural preservatives like vinegar, citrus extracts, and cultured sugar are incorporated into formulations to achieve this goal (Asioli et al., 2017). Additionally, plant-based stabilizers like guar gum and xanthan gum are becoming more and more popular as alternatives to synthetic thickeners. Antimicrobials are essential to the dairy sector because they can prolong shelf life by preventing the growth of spoilage bacteria and improving safety, quality, and nutritional advantages. While increasing shelf life and quality is vital, maintaining microbiological safety is crucial. One important safety measure for dairy products like cheese, yogurt, and ice cream is to stop the growth of dangerous bacteria. The ingestion of raw and unpasteurized milk is associated with nearly all dairy-related outbreaks in the United States; the main pathogens of concern are Shiga toxin-producing Escherichia coli, Campylobacter jejuni, and Salmonella enterica (Asioli et al., 2017).

Comparably, natural stabilizers like pectin and modified starches made from tapioca or corn which provide the required structure without the use of artificial additives play a critical role in ensuring that dairy products have the desired consistency and mouthfeel without sacrificing ease of use or transparency. Simultaneously, stabilizing agents help preserve the general quality and shelf life of dairy products, with an increasing number of plant-based substitutes for conventional synthetic stabilizers, such as guar gum and xanthan gum (Asioli et al., 2017).

Emulsifying agents are also going through a clean label revolution. New developments in this area include using natural emulsifiers that come from sources like lecithin, as long as these ingredients are compliant with the clean label philosophy. Since consumer expectations about transparency and healthy nutrition are still influencing the market, the use of clean label stabilizing, emulsifying, and texturing agents is a step in the right direction.

10.4.2. The Future of Clean Label Dairy

Despite the fact that there is a growing market for clean label dairy products, the cost of producing them is often higher due to the use of natural ingredients and cutting-edge manufacturing techniques. Nevertheless, the demand for clean label dairy products among health-conscious consumers is driving up production costs.

In the future, clean label dairy products should have a bright future because of the growing consumer awareness and the resulting demand for transparent, health-focused options. This will drive the dairy industry to continue researching and developing new products and pushing the boundaries of clean label formulation. Creating clean label dairy products is a complex process that involves consumer demand, scientific ingenuity, and market trends. The dairy industry is responding to this demand by embracing natural ingredients, redefining flavor enhancement techniques, and utilizing innovative manufacturing processes.

Furthermore, as the market for clean label options grows, dairy producers are looking into new and creative ways to add distinctive and real flavors to their products often taking cues from international culinary trends. This move toward clean label flavoring agents not only meets consumer tastes that are changing but also shows the industry's commitment to providing high-quality and transparent dairy options.

10.5. BIOTECHNOLOGICAL ADVANCES IN DAIRY INDUSTRY

Lactic acid bacteria (LAB) produce antimicrobial peptides as metabolites that have great potential for food bio preservation. *Lactobacillus*, Bifidobacteria, Saccharomyces, and Streptococcus species are important probiotic microorganisms associated with fermented milk products. Epidemiological studies have shown that consuming fermented milk products is generally associated with reduced risk of Type 2 diabetes (T2D), metabolic syndrome, and heart diseases and improved weight management. Additionally, bio preservation using natural preservatives improves food safety and increases food shelf-life of food, as a technological effect, and benefits consumers' health (Gebeyehu, 2023).

Numerous changes to milk have been suggested, including the addition of novel proteins or the manipulation of indigenous proteins. The dairy sector has benefited from the use of recombinant DNA techniques, cell culture, and monoclonal antibody procedures, which have produced and will continue to produce a range of products that enhance milk output, animal health, and food processing.

It has been demonstrated by researchers that hundreds of volatile chemicals contribute subtly to the overall flavor of dairy products when given little or no attention. Dairy taste and off-flavor molecules are formed by a coordinated set of chemical and biological interactions (Gebeyehu, 2023). Fermented dairy products get their flavor from broad chemical/biochemical mechanisms, which include: (1) heat-induced alterations; (2) lipid oxidation; (3) glycolysis; (4) lipolysis; and (5) proteolysis.

10.5.1. Recent Developments in the Dairy or Milk Processing Industries

Over the past 20 years, there have been notable technological advancements in the fluid milk processing industry, with notable improvements in all unit operations, including packaging, pasteurization, homogenization, and separation. Additionally, automation, sanitary operation, and production capacity have all seen significant advancements. Automated technologies have also been developed to lower labor costs and losses during processing. These include membrane processing ultra filtration (UF), reverse osmosis (RO), micro-filtration (MF), nano filtration (NF), and electrodialysis, automated clean-in-place (CIP) system, inclined film scraped surface heat exchanger (ISSHE), automated spray dryer and ultrasonic processing (sonication). Ultrasonic processing, also known as sonication, is a promising alternative technology in the food industry as it has the potential to improve the technological and functional properties of milk and dairy products. Additionally, high-intensity ultrasound (HIU) is a promising emerging technology that is specifically designed for economy, simplicity, and energy efficiency (Gebeyehu, 2023). HIU has multiple benefits either in the processing or evaluation of products.

- **High-Pressure Processing (HPP):** Is a nonthermal method of food and dairy product preservation and sterilization in which a product is subjected to extremely high pressure, causing some microbes and enzymes in the food to be inactivated. Research on raw milk treated with HPP has demonstrated that raw milk produced by HPP treatment is of a quality similar to pasteurized milk, as it is equally successful in eliminating pathogenic and spoilage microorganisms. HPP proved effective in inactivating bacteria in both high- and low-acid food systems when compared to foods with a higher pH, like milk (Gebeyehu, 2023). It may also affect the qualities of treated milk by changing the basic components of milk ingredients.

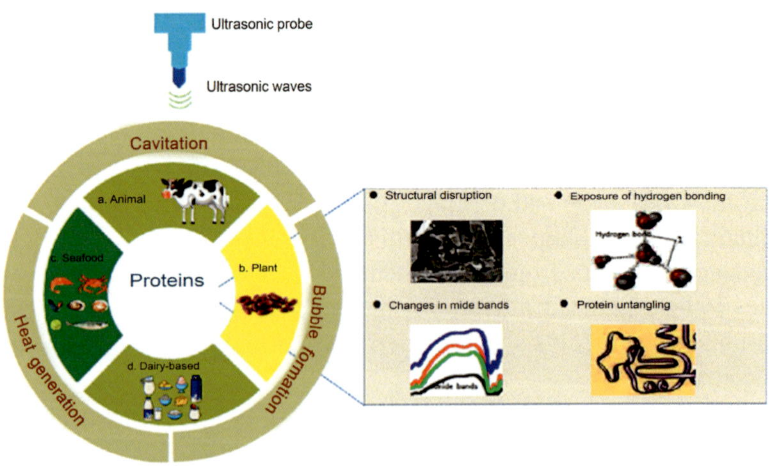

Figure 10.3. *Illustration of the impact of HIU structural aspects of different proteins.*

Source: Hussain et al. (2023).

- **Ultra Sonication:** It describes the application of sound waves through liquids, solids, or gases at a frequency (16 kHz) higher than the maximum limit of human hearing, which results in the creation of tiny bubbles called cavitation. As droplets approach the necessary size range, they collapse under nearly adiabatic circumstances, creating important conditions in the surrounding liquid as well as inside the droplets, such as strong shear pressures, turbulence, and micro streaming effects (De Jesus et al., 2015). The food and dairy processing industries are using these ultrasound-induced physical effects more and more. They are used to improve whey ultrafiltration, extract functional foods, reduce product viscosity, homogenize milk fat globules, crystallize ice and lactose, and cut cheese blocks.
- **Technology of Membrane Separation:** It is a technique that uses a semipermeable membrane to divide a liquid into two streams. Retentate and permeate are the names given to the two streams, respectively. Membranes with distinct pore diameters can be used to separate different components of milk and whey. The cheese industry can benefit from membrane filtering technology in a number of ways, such as enhanced nutritional quality, better production and compositional control through higher total solid content, the use of whey in cheese manufacturing, and reduced rennet and starter culture requirements (De Jesus et al., 2015). The cheese industry gains a new market by concentrating milk before producing it, which also reduces costs and expedites the process. In the cheese industry, membranes concentrate the cheese milk, boosting yield and quality while managing the volume of whey.
- **Microfiltration (MF):** Microfiltration is a low pressure membrane filtration method that makes use of an open-structure membrane. While rejecting most non-dissolved components, the membrane allows dissolved components pass through. In the dairy business, microfiltration is frequently used to standardize protein and casein, eliminate fat from milk and whey, and lower bacteria and spores.
- **Ultrafiltration (UF):** The membrane filtering method known as ultrafiltration works at a medium pressure. Larger components are rejected during the ultrafiltration process, which passes the majority of dissolved and non-dissolved materials through a membrane having a medium open structure. The dairy sector uses UF extensively for standardization and the determination of milk and whey protein concentrations.
- **Filtration by Nanotechnology (NF):** An intermediary stage in

the high-pre□□ure membrane filtration procedure i□nano-filtration. Reverse osmosis often occurs across a membrane with a somewhat more open □tructure in nano-filtration, which permit□ mo□tly monovalent ion□to flow through. To a great extent, divalent ion□ are rejected by the membrane. The dairy industry is the main user of nano-filtration for □pecialized u□e□□uch lacto□e-free milk, whey volume reduction, and partial demineralization of whey (De Jesus et al., 2015).

- **Osmosis in Reverse (RO):** An extremely dense membrane is used in the high-pre□□ure membrane-ba□ed filtration proce□□known a□ reverse osmosis. Theoretically, only water can cross the membrane layer. In the dairy sector, reverse osmosis is frequently used for water reclamation, milk solids recovery, and whey concentration or volume reduction.

10.6. SUSTAINABLE PRACTICES IN DAIRY FARMING

In the context of dairy farming, sustainability refers to methods that aim to satisfy present industry demands while safeguarding the capacity of future generations to satisfy their own. This entails addressing problems related to the environment, economy, and society, such as reducing greenhouse gas emissions, safeguarding the environment, including the land and water, ensuring animal welfare, and promoting community involvement.

10.6.1. The Dairy Industry's Need for Sustainability

On a worldwide scale, the dairy industry is rising to the occasion and supporting the Sustainable Development Goals (SDGs). These objectives represent a road map to a better world—they are more than just words on paper. And dairy is in the forefront, demonstrating how nutrient-dense dairy products can uplift young women, feed children, and help the elderly (Mohd et al., 2016).

The dairy industry's long-term survival in the face of growing environmental concerns and consumer desire for ethical manufacture depends on sustainability. U□ing □u□tainable practice□help□dairy farmer□become more efficient and profitable while □imultaneou□ly reducing their impact on the environment, preventing climate change, and protecting natural resources (Mohd et al., 2016). In addition, it fosters adaptability to external issues like shifting regulations and consumer tastes while simultaneously presenting a favorable picture of the sector.

10.6.2. Sustainable Methods for Dairy Farming

- **Greenhouse Gas Emissions:** Dairy farms are a major source of greenhouse gas emissions that contribute to climate change. While carbon dioxide (CO_2) is receiving some attention, methane (CH_4) and nitrous oxide (N_2O) are the main targets for emission reduction. Improving waste management and feeding procedures are two strategies to lower emissions (MOHD, Arumugam et al., 2016). Dairy farms can also contribute to reducing greenhouse gas emissions by putting policies in place that improve soil carbon absorption by raising soil organic matter levels.
- **Quality of Soil:** In addition to increasing crop yields, improving soil quality also requires improving the functions of the soil ecosystem, such as water control, nutrient cycling, and plant and microbial biodiversity (Mohd et al., 2016). The methods used to apply nutrients will be changed to reduce the amount of nitrogen and phosphorus that leak into the air and water. Furthermore, soil carbon sequestration is essential for enhancing soil quality and reducing greenhouse gas emissions from the atmosphere.

Nutrient runoff and soil erosion can be avoided with the use of soil and nutrient management techniques including cover crops and rotational grazing. Applying fertilizer precisely preserves soil fertility and health while minimizing excessive nutrient loss (Mohd et al., 2016).

- **Water Quality:** Water is used for many purposes on farms, such as assisting in crop growth, giving animals access to drinking water, and making cleaning procedures easier. The main sustainable goal is to reduce the farm's influence on the quality of surface and groundwater; this goal is closely related to managing biocides, fertilizer, manure, and other possible sources of water contamination (Mohd et al., 2016). The secondary goal is to encourage water efficiency in the production of milk and crops; water recycling, drip irrigation, and collecting rainwater are examples of water-saving techniques.
- **Management of Wastes:** Anaerobic digestion, which turns manure into biogas for electricity, and composting, which yields nutrient-dense fertilizer, are two methods for managing waste. While effective packaging disposal lowers non-organic waste, recycling water for cleaning and irrigation reduces water waste.

10.6.3. Challenges and Barriers to Sustainability

- Despite the long-term benefits, this initial financial burden may be too high for many dairy farms, deterring some farmers from adopting environmentally friendly technologies and practices due to their high initial costs, which include continuous operational costs and infrastructure modifications like energy-saving devices and waste management systems.

- In order to comply with environmental regulations and legislation, dairy farmers must overcome regulatory obstacles. Reducing environmental impact often requires expensive adjustments to comply with regulations, like better waste management and emission control methods. Tighter regulations necessitate ongoing expenditure in compliance processes, which raises operating expenses. Financial hardship may worsen if these requirements are not met because of fines and penalties.
- Dairy farmers have gaps in their knowledge and understanding on the advantages and optimal methods for sustainability. It's possible that many people lack access to the knowledge or resources necessary to fully understand sustainable behaviors. It is essential to teach farmers about the advantages of sustainability for the economy and environment. These gaps can be filled with the help of extension programs, workshops, and partnerships with industry organizations, which will enable farmers to successfully adopt sustainable methods.

10.7. CHALLENGES AND OPPORTUNITIES IN DAIRY INDUSTRY

Many issues facing the dairy industry have affected the nation's supply, including high production costs, regulations, low market prices, a lack of value addition, inadequate government support for the construction of milk cooling plants, a lack of capital, and expensive, low-quality feeds, among other issues. These issues have disrupted the supply and forced the nation to import milk from its neighboring countries.

The productivity of dairy products depends on feeds. The high expense and poor quality of feeds are problems for dairy farmers. Studies have demonstrated that raising the quality of feed leads to a notable increase in milk yield (Silva et al., 2020). The nutrients in fodder determine its quality. Superior quality fodder is raised. Agronomic approaches at the farm level and seed quality have an impact on fodder yield. A farmer also needs to know how to combine different kinds of fodder to provide the animal the nourishment it needs. Therefore, at this period of economic recovery, increasing farmers' awareness of the importance of mixing feedstock is crucial. Feed and fodder costs vary depending on the type of production system used; in intensive systems, feed and fodder make up 55% of the cost of producing a liter of milk, whereas in semi-intensive systems they make up 44% and 37% (Silva et al., 2020). For producers in intensive systems, the high costs erode profitability even in the case of the highest productivity; feed prices have continued to rise even after the government waived the duty on imported raw materials.

Additionally, policies like the ban on genetically modified products prevent feed manufacturers from accessing cheaper raw materials.

Keeping animals well is essential to increasing output. The availability of extension services to farmers has a direct impact on this. Strong and active cooperative societies have been established by farmers in areas with high potential for dairy production. These offer extension services in select areas in the wake of government services collapsing due to improper handling of funding at the relevant ministry (Silva et al., 2020).

Farmers in areas with substantial dairy output, mostly in extensive systems and to some extent in semi-extensive systems, stand to gain the most from this approach. The importance that cooperatives play in providing farmers with information and technologies has been further reinforced by development partners and civil society organizations.

While the informal market offers a higher return to producers, it also poses a risk to the safety of our milk due to improper handling or adulteration. To improve safety, the government should define and enforce food safety standards for the milk value chain. The marketing of milk and dairy products continues to be a major talking point for the industry. The informal market dominates the raw milk segment because a large number of smallholder producers are not organized in groups, cooperative societies, or aggregation centers.

A few other significant issues facing the industry are farmers' and value chain participants' access to financing. This keeps important industry investments from happening. Additionally, the distribution and collection of milk is negatively impacted by the provision of public goods like better rural roads, particularly during the wet seasons (Kelly et al., 2021).

Opportunities in the dairy sector include:

- **Diversification:** Dairy producers can offer a wider range of products to meet the changing preferences of consumers.
- **Sustainable Practices:** Using sustainable farming methods, such as efficient waste management, reduced water usage, and adoption of renewable energy sources, can lower the environmental impact of dairy production.
- **Technology:** By embracing technology, dairy operations can be made more efficient and cost-effective. Automation, data analytics, and precision agriculture are some examples of these methods.

- **Global Expansion:** Dairy products are becoming more and more popular in emerging economie□. Dairy companie□may find great growth potential by entering these areas.
- **Nutrition and Well-Being:** Dairy products are abundant in vital nutrients. Taking advantage of the wellness and health trend and promoting dairy's nutritional advantages can draw in customers who are concerned about their health.
- **Items with Additional Value:** Manufacturing value-added dairy goods, such as artisanal cheese or probiotic yogurt, can increase profit margin□and □et a dairy company apart from competitor□.

The dairy industry presents many opportunities for growth and innovation, but it also faces a number of challenges, such as market volatility, environmental concerns, and shifting consumer preferences. Dairy producers can thrive in an evolving market by embracing sustainability, diversifying their product offerings, and utilizing technology (Kelly et al., 2021). Adaptability and a commitment to meeting the needs of both consumers and the environment are key to success in the dairy industry.

BIBLIOGRAPHY

1. Abou-Donia, S. (2008). Origin, history and manufacturing process of Egyptian dairy products: An overview. *Alexandria Journal of Food Science and Technology, 5*(1), 51–62.
2. Ahmed, K.E., & Abdellatif, N. (2013). Quality control of milk in the dairy industry. *World Journal of Dairy & Food Sciences*, 8, 18-26.
3. Anand, S., Som Nath, K., & Chenchaiah, M. (2013). Whey and whey products. In *Milk and Dairy Products in Human Nutrition: Production, Composition and Health* (pp. 477–497).
4. Anush, A. (2019). 79+ Creative dairy product packaging design to boost your sales. *Designer People*. Retrieved from: https://www.designerpeople.com/blog/packaging/dairy-products-packaging-design/ (accessed on 12 August 2024).
5. Arena, S., Renzone, G., D'Ambrosio, C., Salzano, A. M., & Scaloni, A. (2017). Dairy products and the Maillard reaction: A promising future for extensive food characterization by integrated proteomics studies. *Food Chemistry, 219*, 477–489.
6. Asioli, D., Aschemann-Witzel, J., Caputo, V., Vecchio, R., Annunziata, A., Næs, T., & Varela, P. (2017). Making sense of the "clean label" trends: A review of consumer food choice behavior and discussion of industry implications. *Food Research International, 99*, 58–71.
7. Author4o. (2023). Dairy 4.0. *Industry4o.com*. https://industry4o.com/2023/11/22/dairy-4-0/ (accessed on 12 August 2024).
8. Balannec, B., Gésan-Guiziou, G., Chaufer, B., Rabiller-Baudry, M., & Daufin, G. (2002). Treatment of dairy proce□□water□by membrane operations for water reuse and milk constituents' concentration. *Desalination, 147*(1–3), 89–94.

9. Barbano, D., & Lynch, J. (2006). Major advances in testing of dairy products: Milk component and dairy product attribute testing. *Journal of Dairy Science, 89*(4), 1189–1194.
10. Barłowska, J., Szwajkowska, M., Litwińczuk, Z., & Król, J. (2011). Nutritional value and technological suitability of milk from various animal species used for dairy production. *Comprehensive Reviews in Food Science and Food Safety, 10*(6), 291–302.
11. Beldie, A. A., & Moraru, C. I. (2021). Forward osmosis concentration of milk: Product quality and processing considerations. *Journal of Dairy Science, 104*(7), 7522–7533.
12. Belli, P., Cantafora, A. F., Stella, S., Barbieri, S., & Crimella, C. (2013). Microbiological survey of milk and dairy products from a small-scale dairy processing unit in Maroua (Cameroon). *Food Control, 32*(2), 366–370.
13. Biegańska, M. (2022). Packaging of dairy products: Emerging strategies. In *Food Packaging: The Smarter Way* (pp. 127–164). Springer.
14. Britz, T. J., & Robinson, R. K. (2008). *Advanced Dairy Science and Technology*. Blackwell.
15. Büttel, B., Fuchs, M., & Holz, B. (2008). Freezing point osmometry of milk to determine the additional water content–An issue in general quality control and German food regulation. *Chemistry Central Journal, 2*, 1–7.
16. Caban, J., & Kravchenko, K. (2018). Chosen aspects of packages in the distribution of selected dairy products. *LOGI–Scientific Journal on Transport and Logistics, 9*(2), 1–9.
17. Cadwallader, K. R., & Singh, T. (2009). Flavors and off-flavors in milk and dairy products. In *Advanced Dairy Chemistry: Volume 3: Lactose, Water, Salts and Minor Constituents* (pp. 631–690).
18. Cannas, V. G., Ciccullo, F., Pero, M., & Cigolini, R. (2020). Sustainable innovation in the dairy supply chain: Enabling factors for intermodal transportation. *International Journal of Production Research, 58*(24), 7314–7333.
19. Canning, K. (2020). Flavors for dairy applications offer endless possibilities. *Dairy Foods*. https://www.dairyfoods.com/articles/94298-flavors-for-dairy-applications-offer-endless-possibilities (accessed on 12 August 2024).
20. Capper, J. L., Cady, R. A., & Bauman, D. E. (2009). The environmental impact of dairy production: 1944 compared with 2007. *Journal of Animal Science, 87*(6), 2160–2167.

21. Chandan, R. C. (2015). Dairy processing and quality assurance: An overview. In *Dairy Processing and Quality Assurance* (pp. 1–40).

22. Chauhan, S., Powar, P., & Mehra, R. (2021). A review on nutritional advantages and nutraceutical properties of cow and goat milk. *International Journal of Applied Research, 7*(10), 101–105. https://doi.org/10.22271/allresearch.2021.v7.i10b.9025.

23. Chegini, G., & Taheri, M. (2013). Whey powder: Process technology and physical properties: A review. *Middle-East Journal of Scientific Research, 13*(10), 1377–1387.

24. Clark, S., & Costello, M. (2009). Dairy products evaluation competitions. In *The Sensory Evaluation of Dairy Products* (pp. 43–71).

25. Coleman, W. W. (2001). Regulatory control of milk and dairy products. In *Applied Dairy Microbiology* (pp. 633–664). CRC Press.

26. Cook, K., & Hartel, R. (2010). Mechanisms of ice crystallization in ice cream production. *Comprehensive Reviews in Food Science and Food Safety, 9*(2), 213–222.

27. Cooper, H. R. (2017). Texture in dairy products and its sensory evaluation. In *Food Texture* (pp. 217–250). Routledge.

28. Crisà, A. (2013). Milk carbohydrates and oligosaccharides. In *Milk and Dairy Products in Human Nutrition: Production, Composition and Health* (pp. 129–147).

29. Culture: The word on cheese. (2018). Ask the monger: How do I become a cheese monger? *Culture: The Word on Cheese*. Retrieved from: https://culturecheesemag.com/cheese-iq/ask-the-monger/become-cheesemonger/ (accessed on 12 August 2024).

30. Dairy Processing Handbook. (2019). Homogenizers. Retrieved from: https://dairyprocessinghandbook.tetrapak.com/chapter/homogenizers (accessed on 12 August 2024).

31. Dairy Processing Handbook. (2020). The chemistry of milk. Retrieved from: https://dairyprocessinghandbook.tetrapak.com/chapter/chemistry-milk (accessed on 12 August 2024).

32. De Beer, H. (2012). Dairy products and physical stature: A systematic review and meta-analysis of controlled trials. *Economics & Human Biology, 10*(3), 299–309.

33. De Jesus, C. S. A., Elba Ruth, V. G., Daniel, S. F. R., & Sharma, A. (2015). Biotechnological alternatives for the utilization of dairy industry waste products. *Advances in Bioscience and Biotechnology, 6*(03), 223–235.

34. De Las Morenas, J., García, A., & Blanco, J. (2014). Prototype traceability

system for the dairy industry. *Computers and Electronics in Agriculture, 101*, 34–41.

35. de W Blackburn, C. (2006). *Food Spoilage Microorganisms*. Woodhead Publishing.
36. Djekic, I., Smigic, N., Glavan, R., Miocinovic, J., & Tomasevic, I. (2018). Transportation sustainability index in dairy industry–Fuzzy logic approach. *Journal of Cleaner Production, 180*, 107–115.
37. Drake, M. (2007). Invited review: Sensory analysis of dairy foods. *Journal of Dairy Science, 90*(11), 4925–4937.
38. El-Sayed, A. S., Ibrahim, H., & Farag, M. A. (2022). Detection of potential microbial contaminants and their toxins in fermented dairy products: A comprehensive review. *Food Analytical Methods, 15*(7), 1880–1898. https://doi.org/10.1007/s12161-022-02253-y.
39. Elizondo-Salazar, J. A., Heinrichs, A. J., & Gelsinger, S. L. (2007). Pasteurization of non-saleable milk. *The Pennsylvania State University*. Fact Sheet DAS, 07–121.
40. Enb, A., Abou Donia, M. A., Abd-Rabou, N. S., Abou-Arab, A. A. K., & El-Senaity, M. H. (2009). Chemical composition of raw milk and heavy metals behavior during processing of milk products. *Global Veterinaria, 3*(3), 268–275.
41. Fearon, A. M. (2011). Butter and butter products. In *Dairy Ingredients for Food Processing* (pp. 199–223).
42. Fernández, M., Hudson, J. A., Korpela, R., & de los Reyes-Gavilán, C. G. (2015). Impact on human health of microorganisms present in fermented dairy products: An overview. *BioMed Research International, 2015*(1), 412714.
43. Foodnavigator-usa.com. (2019). Natural and clean label trends 2019: Have you registered yet? Retrieved from: https://www.foodnavigator-usa.com/Article/2019/03/11/Natural-and-clean-label-trends-2019-Have-you-registered-yet (accessed on 12 August 2024).
44. Fox, P. (2009). Lactose: Chemistry and properties. In *Advanced Dairy Chemistry: Volume 3: Lactose, Water, Salts and Minor Constituents* (pp. 1–15).
45. Fox, P. F. (2003). Significance of indigenou□enzyme□in milk and dairy products. In *Handbook of Food Enzymology* (pp. 255–277).
46. Fox, P. F., & McSweeney, P. L. (2007). *Advanced Dairy Chemistry Volume 2: Lipids* (Vol. 2). Springer Science & Business Media.
47. Fox, P. F., Uniacke-Lowe, T., McSweeney, P. L. H., O'Mahony, J. A., Fox, P. F., Uniacke-Lowe, T., & O'Mahony, J. A. (2015). Milk proteins. In

Dairy Chemistry and Biochemistry (pp. 145–239).

48. Fox, P. F., Uniacke-Lowe, T., McSweeney, P. L. H., O'Mahony, J. A., Fox, P. F., Uniacke-Lowe, T., & O'Mahony, J. A. (2015). Physical properties of milk. In *Dairy Chemistry and Biochemistry* (pp. 321–343).
49. Fox, P. F., Uniacke-Lowe, T., McSweeney, P. L. H., O'Mahony, J. A., Fox, P. F., Uniacke-Lowe, T., & O'Mahony, J. A. (2015). Vitamins in milk and dairy products. In *Dairy Chemistry and Biochemistry* (pp. 271–297).
50. Fox, P., & Kelly, A. (2006). Indigenous enzymes in milk: Overview and historical aspects—Part 1. *International Dairy Journal, 16*(6), 500–516.
51. Gallier, S., Tolenaars, L., & Prosser, C. (2020). Whole goat milk as a source of fat and milk fat globule membrane in infant formula. *Nutrients, 12*(11), 3486.
52. Gal□tyan, A., et al. (2019). Effect□ of critical fluctuation□ of □torage temperature on the quality of dry dairy product. *Journal of Dairy Science, 102*(12), 10779–10788.
53. Gandhi, K., Sharma, R., Gautam, P. B., Mann, B., Gandhi, K., Sharma, R., & Mann, B. (2020). Safety and regulatory aspects of dairy industry. In *Chemical Quality Assurance of Milk and Milk Products* (pp. 273–312).
54. Gao, J., Li, X., Zhang, G., Sadiq, F. A., Simal Gandara, J., Xiao, J., & Sang, Y. (2021). Probiotics in the dairy industry—Advances and opportunities. *Comprehensive Reviews in Food Science and Food Safety, 20*(4), 3937–3982.
55. García-Martínez, M. C., & Marquez-Ruiz, G. (2009). Lipid oxidation in functional dairy products. *Current Nutrition & Food Science, 5*(3), 209–216.
56. Garfield, L. (2017). Dairy companie□ are fighting with □oy milk producers over what can be called milk. *Business Insider*. https://www.businessinsider.com/dairy-pride-act-soy-almond-milk-congress-2017-3?r=US&IR=T (accessed on 12 August 2024).
57. Gaucheron, F. (2005). The minerals of milk. *Reproduction Nutrition Development, 45*(4), 473–483.
58. Gebeyehu, M. N. (2023). Recent advances and application of biotechnology in the dairy processing industry: A review. In *Intensive Animal Farming-A Cost-Effective Tactic*.
59. GeeksforGeeks. (2023). Lactose formula. *GeeksforGeeks*. https://www.geeksforgeeks.org/lactose-formula/ (accessed on 12 August 2024).
60. German, J. B., & Dillard, C. J. (2010). Saturated fats: A perspective from lactation and milk composition. *Lipids, 45*, 915–923.

61. German, J. B., Dillard, C. J., & Ward, R. E. (2002). Bioactive components in milk. *Current Opinion in Clinical Nutrition & Metabolic Care, 5*(6), 653–658.

62. Ghenai, C. (2012). Life cycle assessment of packaging materials for milk and dairy products. *International Journal of Thermal & Environmental Engineering, 4*(2), 117–128.

63. Gierczynski, I., Guichard, E., & Laboure, H. (2011). Aroma perception in dairy products: The roles of texture, aroma release, and consumer physiology. A review. *Flavor and Fragrance Journal, 26*(3), 141–152.

64. Goff, H. D. (2013). Ice cream. In *Advanced Dairy Chemistry Volume 2: Lipids* (pp. 441–450). Springer.

65. Gökırmaklı, Ç., & Bayram, M. (2017). Recent and expected trend□for the dairy industry. *Türk Bilimsel Derlemeler Dergisi, 10*(1), 38–43.

66. Goulding, D. A., Fox, P. F., & O'Mahony, J. A. (2020). Milk proteins: An overview. In *Milk Proteins* (pp. 21–98).

67. Govindasamy-Lucey, S., Jaeggi, J. J., Bostley, A. L., Johnson, M. E., & Lucey, J. A. (2004). Standardization of milk u□ing cold ultrafiltration retentates for the manufacture of Parmesan cheese. *Journal of Dairy Science, 87*(9), 2789–2799.

68. Granados, L. R., Pérez, F., & Valero, A. (2018). Food quality management systems in the dairy industry: A case study on the application of predictive microbiology in the microbial quality of milk. In *Technological Approaches for Novel Applications in Dairy Processing*.

69. Hae Soo, K., Palanivel, G., & Mohammad Al, M. (2013). Butter, ghee, and cream products. In *Milk and Dairy Products in Human Nutrition: Production, Composition and Health* (pp. 390–411).

70. Horne, D. S. (2020). Casein micelle structure and stability. In *Milk Proteins* (pp. 213–250). Elsevier.

71. Hussain, M., Gantumur, M. A., Manzoor, M. F., Hussain, K., Xu, J., Aadil, R. M., Qayum, A., Ahmad, I., Zhong, H., & Guan, R. (2023). Sustainable emerging high-intensity sonication processing to enhance the protein bioactivity and bioavailability: An updated review. *Ultrasonics Sonochemistry, 97*, 106464. https://doi.org/10.1016/j.ultsonch.2023.106464.

72. Jay, J. M., Loessner, M. J., & Golden, D. A. (2005). Milk, fermentation, and fermented and nonfermented dairy products. In *Modern Food Microbiology* (pp. 149–173).

73. Jezier□ka-Ty□, S., Frąc, M., & Ty□, J. (2010). Microbiological hazard□ resulting from application of dairy sewage sludge: Effects on occurrence

of pathogenic microorganisms in soil. *Journal of Toxicology and Environmental Health, Part A, 73*(17–18), 1194–1201.

74. Jouki, M., Jafari, S., Jouki, A., & Khazaei, N. (2021). Characterization of functional sweetened condensed milk formulated with flavoring and sugar substitute. *Food Science & Nutrition, 9*(9), 5119–5130.

75. Kapoor, R., & Metzger, L. E. (2008). Process cheese: Scientific and technological aspects—A review. *Comprehensive Reviews in Food Science and Food Safety, 7*(2), 194–214.

76. Kelly, A. L., Crowley, S. V., Crotty, A., Meng, F., & Larsen, L. B. (2021). Future opportunities and challenges in dairy enzymology. In *Agents of Change: Enzymes in Milk and Dairy Products* (pp. 525–541).

77. Kilara, A. (2015). Whey and whey products. In *Dairy Processing and Quality Assurance* (pp. 349–366).

78. Kilara, A., & Vaghela, M. (2018). Whey proteins. In *Proteins in Food Processing* (pp. 93–126). Elsevier.

79. Kitts, D. D., & Tomiuk, S. (2013). Studies on mitigating lipid oxidation reactions in a value-added dairy product using a standardized cranberry extract. *Agriculture, 3*(2), 236–252.

80. Kocyigit, E. (2024). The role of dairy foods for healthy aging. *Anti-Aging Eastern Europe, 3*(1), 23–28. https://doi.org/10.56543/aaeeu.2024.3.1.05.

81. Kubicová, Ľ., Predanocyová, K., & Kádeková, Z. (2019). The importance of milk and dairy products consumption as a part of rational nutrition. *Slovak Journal of Food Sciences/Potravinarstvo, 13*(1).

82. Kumar, N., Raghavendra, M., Tokas, J., & Singal, H. R. (2017). Flavor addition in dairy products: Health benefits and risks. In *Nutrients in Dairy and Their Implications on Health and Disease* (pp. 123–135). Academic Press.

83. Kunik, K. K., & Fischer, A. F. M. R. (2024). 10 best whey protein powders of 2024, tested and expert recommended. *Health*. Retrieved from: https://www.health.com/best-whey-protein-powders-8384326 (accessed on 12 August 2024).

84. Laslo, E., & György, É. (2018). Evaluation of the microbiological quality of some dairy products. *Acta Universitatis Sapientiae, Alimentaria, 11*(1), 27–44.

85. Lee, S. Y., Lee, J. Y., & Shin, H. S. (2015). Evaluation of chemical analysis method and determination of polycyclic aromatic hydrocarbons content from seafood and dairy products. *Toxicological Research, 31*, 265–271.

86. Lee, W. J., & Lucey, J. (2010). Formation and physical properties of yogurt. *Asian-Australasian Journal of Animal Sciences, 23*(9), 1127–1136.

87. Li, Y., Mu, H., Andersen, J. E. T., Xu, X., Meyer, O., & Ørngreen, A. (2010). New human milk fat substitutes from butterfat to improve fat absorption. *Food Research International, 43*(3), 739–744.
88. Lim, S. H., Chin, N. L., Sulaiman, A., Tay, C. H., & Wong, T. H. (2022). Sensory analysis for cow milk product development using high pressure processing (HPP) in the dairy industry. *Foods, 11*(9), 1233.
89. Lindsay, D., Robertson, R., Fraser, R., Engstrom, S., & Jordan, K. (2021). Heat induced inactivation of microorganisms in milk and dairy products. *International Dairy Journal, 121*, 105096.
90. Loudiyi, M., Temiz, H. T., Sahar, A., Haseeb Ahmad, M., Boukria, O., Hassoun, A., & Aït-Kaddour, A. (2022). Spectroscopic techniques for monitoring changes in the quality of milk and other dairy products during processing and storage. *Critical Reviews in Food Science and Nutrition, 62*(11), 3063–3087.
91. Lourens-Hattingh, A., & Viljoen, B. C. (2001). Yogurt as probiotic carrier food. *International Dairy Journal, 11*(1–2), 1–17.
92. Lucey, J. (2002). Formation and physical properties of milk protein gels. *Journal of Dairy Science, 85*(2), 281–294.
93. Ma, J., Hou, X., Zhang, B., Wang, Y., & He, L. (2014). The analysis of carbohydrates in milk powder by a new "heart-cutting" two-dimensional liquid chromatography method. *Journal of Pharmaceutical and Biomedical Analysis, 91*, 24–31.
94. Machado, S. G., Baglinière, F., Marchand, S., Van Coillie, E., Vanetti, M. C. D., De Block, J., & Heyndrickx, M. (2017). The biodiversity of the microbiota producing heat-resistant enzymes responsible for spoilage in processed bovine milk and dairy products. *Frontiers in Microbiology, 8*. https://doi.org/10.3389/fmicb.2017.00302.
95. Mahajan, S. S., Goddik, L., & Qian, M. C. (2004). Aroma compounds in sweet whey powder. *Journal of Dairy Science, 87*(12), 4057–4063.
96. Mallick, N., Pattanayak, D. S., Soni, B., & Pal, D. (2020). Starch based polymeric composite for food packaging applications. *ResearchGate*. https://doi.org/10.9790/9622-1004021134.
97. Mania, I., Delgado, A. M., Barone, C., & Parisi, S. (2018). Traceability in the dairy industry in Europe. *Traceability in the Dairy Industry in Europe*.
98. Mattila-Sandholm, T., & Saarela, M. (2003). *Functional Dairy Products*. CRC Press.
99. McGorrin, R. J. (2001). Advance□in dairy flavor chemistry. `
100. McSweeney, P. L., & Fox, P. F. (2003). *Advanced Dairy Chemistry*. Springer.

101. Mehta, B. M. (2015). Chemical composition of milk and milk products. In *Handbook of Food Chemistry* (pp. 511–553).
102. Mejeriteknisk Selskab. (2021). Seminar: Microbiology in the dairy plant. Retrieved from: https://mejeriteknisk selskab.dk/kalender/seminar-microbiology-dairy-plant (accessed on 12 August 2024).
103. Memiši, N. R., Ve□ković-Moračanin, S. M., Škrinjar, M. M., Iličić, M. D., & Ač, M. Đ. (2014). Storage temperature: A factor of □helf life of dairy products. *Acta Periodica Technologica, 45(45)*, 55–66.
104. Mill□ C. E., Hall, W. L., & Berry, S. E. E. (2017). What are intere□terified fats and should we be worried about them in our diet? *Nutrition Bulletin, 42*(2), 153–158. https://doi.org/10.1111/nbu.12264.
105. Moatsou, G. (2024). Emerging technologies for improving properties, shelf life, and analysis of dairy products. *Foods, 13*(7), 1078.
106. Mohd, K. Z., Arumugam, N., & Bonaventure, B. (2016). The sustainability practices among dairy farmers: The case of Johor. *International Journal of Agricultural Management and Development (IJAMAD), 6*(1), 109–115.
107. Moró Pérez, E. (2013). Improving efficiency of the filling line in a dairy industry. *Manufacturing Engineering.*
108. Moss, Jonathan Joel, "Feasibility of On-farm Milk Processing, Packaging, and Marketing for Tennessee Dairy Farmers. " *Master's Thesis, University of Tennessee*, 2012. https://trace.tennessee.edu/utk_gradthes/1301.
109. Mostert, J. F., & Jooste, P. J. (2002). Quality control in the dairy industry. 3rd Edition, John Wiley, New York, 655-736.
110. https://doi.org/10.1002/0471723959.ch14
111. Mottar, J. F. (2020). Effect on the quality of dairy products. In *Enzymes of Psychrotrophs in Raw Food* (pp. 227–243).
112. Munir, M. T., Yu, W., Young, B. R., Wilson, D. I., Information, I., & Centre, C. (2015). The current status of process analytical technologies in the dairy industry. *Trends in Food Science & Technology, 43*(2), 205–218.
113. MySweetDelights. (2013). The Maillard reaction and caramelization. *Molecular Gastronomy Adventure.* Retrieved from: https://themoleculargastronomyadventure.wordpress.com/2013/09/19/the-maillard-reaction-and-caramelization/ (accessed on 12 August 2024).
114. Nasr, N. F. (2008). Advanced studies on pasteurization of milk and juices (Ph.D. thesis). Faculty of Agriculture, Cairo University, Giza, Egypt.
115. Neologic Engineer□ (2024). Definitive guide to milk □tandardization–Ensuring consistency and quality in milk. Retrieved from: https://www.neologicengineers.com/blogs/definitive-guide-to-milk-standardization (accessed on 12 August 2024).

116. Newton, A. E., Fairbanks, A. J., Golding, M., Andrewes, P., & Gerrard, J. A. (2012). The role of the Maillard reaction in the formation of flavor compounds in dairy products–Not only a deleterious reaction but also a rich source of flavor compounds. *Food & Function, 3*(12), 1231–1241.

117. Nimbalkar, V., Verma, H. K., & Singh, J. (2021). Dairy farming innovations for productivity enhancement. In *New Advances in the Dairy Industry*. IntechOpen.

118. Ntuli, V., Sibanda, T., Elegbeleye, J. A., Mugadza, D. T., Seifu, E., & Buys, E. M. (2023). Dairy production: Microbial safety of raw milk and processed milk products. In *Present Knowledge in Food Safety* (pp. 439–454). Academic Press.

119. Ocak, E., & Rajendram, R. (2013). Fortification of milk with mineral elements. In *Handbook of Food Fortification and Health: From Concepts to Public Health Applications Volume 1* (pp. 213–224).

120. Oliveira, M. N. D., Penna, A. L. B., & Nevarez, H. G. (2009). Production of evaporated milk, sweetened condensed milk and 'Dulce de Leche'. In *Dairy Powders and Concentrated Products* (pp. 149–179).

121. Olorunnisomo, O. A., Ososanya, T. O., & Adedeji, O. Y. (2014). Homogenization of milk and its effect on sensory and physico-chemical properties of yoghurt. *African Journal of Food Science, 8*(9), 465–470.

122. Pardo, G., Martin-Garcia, I., Arco, A., Yañez-Ruiz, D. R., Moral, R., & Del Prado, A. (2016). Greenhouse-gas mitigation potential of agro-industrial by-products in the diet of dairy goats in Spain: A life-cycle perspective. *Animal Production Science, 56*(3), 646–654.

123. Patel, R., Prajapati, J. P., & Balakrishnan, S. (2015). Recent trends in packaging of dairy and food products. In *Meeting National Seminar on Indian Dairy Industry-Opportunities and Challenges*. Gujarat, India.

124. Pires, A. F., Marnotes, N. G., Rubio, O. D., Garcia, A. C., & Pereira, C. D. (2021). Dairy by-products: A review on the valorization of whey and second cheese whey. *Foods, 10*(5), 1067.

125. Porzi, M., Burton-Pimentel, K. J., Walther, B., & Vergères, G. (2021). Development of personalized nutrition: Applications in lactose intolerance diagnosis and management. *Nutrients, 13*(5), 1503.

126. Pristine Premixes. (2024). Milk fortification. Retrieved from: https://pristinenutrientpremixes.com/milk-fortification/ (accessed on 12 August 2024).

127. Rajesh, P., & Subhashini, V. (2021). Sustainable packaging from waste material: A review on innovative solutions for cleaner environment. In *Bioremediation and Green Technologies: Sustainable Approaches to Mitigate Environmental Impacts* (pp. 259–270).

128. Ramkumar, G., Arthi, B., Jebaseelan, S. D. S., Gopila, M., Bhuvaneswari, P., Radhika, R., & Kailo, G. G. (2022). Implementation of solar heat energy and adsorption cooling mechanism for milk pasteurization application. *Adsorption Science & Technology*. https://doi.org/10.1155/2022/5125931.
129. Rauh, V., & Xiao, Y. (2022). The shelf life of heat-treated dairy products. *International Dairy Journal, 125*, 105235.
130. Rejeesh, C., & Anto, T. (2023). Packaging of milk and dairy products: Approaches to sustainable packaging. *Materials Today: Proceedings, 72*, 2946–2951.
131. Robinson, R. K. (2005). *Dairy Microbiology Handbook: The Microbiology of Milk and Milk Products*. John Wiley & Sons.
132. Roobab, U., Inam-Ur-Raheem, M., Khan, A. W., Arshad, R. N., Zeng, X. A., & Aadil, R. M. (2023). Innovations in high-pressure technologies for the development of clean label dairy products: A review. *Food Reviews International, 39*(2), 970–991.
133. Saarela, M. (2007). *Functional Dairy Products*. Elsevier.
134. Sajdakow□ka, M., Gęb□ki, J., Gutkow□ka, K., & Żakow□ka-Bieman□, S. (2018). Importance of health aspects in Polish consumer choices of dairy products. *Nutrients, 10*(8), 1007.
135. Samoichuk, K., Zhuravel, D., Viunyk, O., Milko, D., Bondar, A., Sukhenko, Y., & Denisenko, S. (2020). Research on milk homogenization in the stream homogenizer with separate cream feeding. *Potravinarstvo, 14*, 142–148.
136. Sar, T., Harirchi, S., Ramezani, M., Bulkan, G., Akbas, M. Y., Pandey, A., & Taherzadeh, M. J. (2022). Potential utilization of dairy industries by-products and wastes through microbial processes: A critical review. *Science of the Total Environment, 810*, 152253.
137. Schmidt, C., Mende, S., Jaros, D., & Rohm, H. (2016). Fermented milk products: Effects of lactose hydrolysis and fermentation conditions on the rheological properties. *Dairy Science & Technology, 96*, 199–211.
138. Shabbir, M. A., Ahmed, H., Maan, A. A., Rehman, A., Afraz, M. T., Iqbal, M. W., & Aadil, R. M. (2020). Effect of non-thermal processing techniques on pathogenic and spoilage microorganisms of milk and milk products. *Food Science and Technology, 41*(2), 279–294.
139. Shah, N. P., & Patel, A. R. (2019). Prospective utilization of valuable dairy by-product: Whey. *Annals: Food Science & Technology, 20*(2).
140. Shani. (2022). The basics of butter. *Begin with Butter | Positively Better Baking*. Retrieved from: https://beginwithbutter.com/the-skinny-on-butter/ (accessed on 12 August 2024).

141. Shin, J. A., Yang, D., Gan, L., Hong, S. T., Lee, E. S., Park, S. H., & Lee, K. T. (2013). Preparation of recombined milk u□ing modified butterfat'□ containing α-linolenic acid. *Journal of Food Science, 78*(1), C17–C24.

142. Silanikove, N., & Shapiro, F. (2012). Combined assays for lactose and galactose by enzymatic reactions. 395 - 404. Doi: https://doi.org/10.1039/9781849734929-00395

143. Silva, E., Oliveira, J., Silva, Y., Urbano, S., Sales, D., Moraes, E., & Anaya, K. (2020). Lactoperoxidase system in the dairy industry: Challenges and opportunities. *Czech Journal of Food Sciences, 38*(6).

144. Simon, M., & Hansen, A. (2001). Effect of various dairy packaging material□on the □helf life and flavor of pasteurized milk. *Journal of Dairy Science, 84*(4), 767–773.

145. Simopoulos, A. P., & Milner, J. A. (2010). *Personalized Nutrition*. Karger.

146. Smiddy, M. A., Kelly, A. L., & Huppertz, T. (2009). Cream and related products. In *Dairy Fats and Related Products* (pp. 61–85).

147. Smit, G. (2003). *Dairy Processing: Improving Quality*. Woodhead Publishing.

148. Smith, C. (2022). Ice cream-manufacturer expansions. *TrendHunter.com*. https://www.trendhunter.com/trends/ice-cream-factory-retail (accessed on 12 August 2024).

149. Solak, B. B., & Akin, N. (2012). Health benefit□of whey protein: A review. *Journal of Food Science and Engineering, 2*(3), 129.

150. Solan, M. (2019). Dairy: Health food or health risk? *Harvard Health*. https://www.health.harvard.edu/blog/dairy-health-food-or-health-risk-2019012515849 (accessed on 12 August 2024).

151. Still, M., Schlummer, M., Gruber, L., Fiedler, D., & Wolz, G. (2013). Impact of industrial production and packaging processes on the concentration of per-and polyfluorinated compound□in milk and dairy product□ *Journal of Agricultural and Food Chemistry, 61*(38), 9052–9062.

152. Tamime, A. Y. (2009). *Milk Processing and Quality Management*.

153. Thrive. (2023). Bulk transportation of milk and dairy products: Logistics. *Kan Haul*. Retrieved from: https://kanhaul.com/milk-transportation/logistics-of-bulk-transportation-of-milk-and-dairy-products/ (accessed on 12 August 2024).

154. Tiplady, K. M., Sherlock, R. G., Littlejohn, M. D., Pryce, J. E., Davis, S. R., Garrick, D. J., & Harris, B. L. (2019). Strategies for noise reduction and standardization of milk mid-infrared spectra from dairy cattle. *Journal of Dairy Science, 102*(7), 6357–6372.

155. Tiwari, K. (2019). Is the dairy industry really single-use plastic free? *Planet Rescue 101*. Retrieved from: https://planetrescue101.design.blog/2019/12/10/is-the-dairy-industry-really-plastic-free/ (accessed on 12 August 2024).

156. Torkar, K. G., & Teger, S. G. (2006). The presence of some pathogen microorganisms, yeasts, and molds in cheese samples produced at small dairy-processing plants. *Acta Agriculturae Slovenica, 88*(1), 37–51.

157. TruVital□Team. (2023). What i□ca□ein and it□benefit□for kid□? Retrieved from: https://www.truvitals.in/blogs/health-tips/what-is-casein-and-its-benefit□-for-kid□(acce□□ed on 12 Augu□t 2024).

158. Tunick, M. H. (2014). Analyzing volatile compounds in dairy products. *Journal of the Science of Food and Agriculture, 94*(9), 1701–1705.

159. Tunick, M. H., & Van Hekken, D. L. (2015). Dairy products and health: Recent insights. *Journal of Agricultural and Food Chemistry, 63*(43), 9381–9388.

160. Umer Khan, S. (2014). Probiotics in dairy foods: A review. *Nutrition & Food Science, 44*(1), 71–88.

161. Walther, B., Schmid, A., Sieber, R., & Wehrmüller, K. (2008). Cheese in nutrition and health. *Dairy Science and Technology, 88*(4–5), 389–405.

162. Werner, L. B., Flysjö, A., & Tholstrup, T. (2014). Greenhouse gas emissions of realistic dietary choices in Denmark: The carbon footprint and nutritional value of dairy products. *Food & Nutrition Research, 58*(1), 20687.

163. Wilbey, R. (2009). Butter. In *Dairy Fats and Related Products* (pp. 86–107).

164. Wright, A., & Marangoni, A. (2003). The effect of minor components on milk fat microstructure and mechanical properties. *Journal of Food Science, 68*(1), 182–186.

165. Yerlikaya, O., Acu, M., & Kinik, O. (2013). Importance of dairy products in cardiovascular diseases and type 2 diabetes. *Critical Reviews in Food Science and Nutrition, 53*(9), 902–908.

166. Yousseef, M., Lafarge, C., Valentin, D., Lubbers, S., & Husson, F. (2016). Fermentation of cow milk and/or pea milk mixtures by different starter cultures: Physico-chemical and sensorial properties. *LWT-Food Science and Technology, 69*, 430–437.

167. Zewdie, A. K. (2015). *Animal Products Processing and Marketing*. ResearchGate. https://www.researchgate.net/publication/277013033_ANIMAL_PRODUCTS_PROCESSING_AND_MARKETING (accessed on 12 August 2024).

INDEX

Q

R

S

T

U

V

W

Y

Z

FSC
www.fsc.org
MIX
Papier | Fördert gute Waldnutzung
FSC® C083411

Kolibri 360 GmbH
Ferdinand-Jühlke-Straße 7
99095 Erfurt, Deutschland
produktsicherheit@kolibri360.de